THE DANGERS OF AUTOMATION IN AIRLINERS

ACCIDENTS WAITING TO HAPPEN

Jack J Hersch

AIR WORLD

AIR WORLD

THE DANGERS OF AUTOMATION IN AIRLINERS
Accidents Waiting to Happen

First published in Great Britain in 2020 by
Air World
An imprint of
Pen & Sword Books Ltd
Yorkshire – Philadelphia

Copyright © Jack J Hersch, 2020

ISBN 978 1 52677 314 2

The right of Jack J Hersch to be identified as Author of this work has been asserted by him in accordance with the Copyright, Designs and Patents Act 1988.

A CIP catalogue record for this book is available from the British Library.

All rights reserved. No part of this book may be reproduced or transmitted in any form or by any means, electronic or mechanical including photocopying, recording or by any information storage and retrieval system, without permission from the Publisher in writing.

Typeset by SJmagic DESIGN SERVICES, India.

Printed and bound in the UK by TJ Books Ltd.

Pen & Sword Books Limited incorporates the imprints of Atlas, Archaeology, Aviation, Discovery, Family History, Fiction, History, Maritime, Military, Military Classics, Politics, Select, Transport, True Crime, Air World, Frontline Publishing, Leo Cooper, Remember When, Seaforth Publishing, The Praetorian Press, Wharncliffe Local History, Wharncliffe Transport, Wharncliffe True Crime and White Owl.

For a complete list of Pen & Sword titles please contact

PEN & SWORD BOOKS LIMITED
47 Church Street, Barnsley, South Yorkshire, S70 2AS, England
E-mail: enquiries@pen-and-sword.co.uk
Website: www.pen-and-sword.co.uk

Or
PEN AND SWORD BOOKS
1950 Lawrence Rd, Havertown, PA 19083, USA
E-mail: Uspen-and-sword@casematepublishers.com
Website: www.penandswordbooks.com

MIX
Paper from
responsible sources
FSC
www.fsc.org   FSC® C013056

Contents

Part III: Landing Gear Down

Part IV: MAX

Part V: Landing Gear Up

CONTENTS

Figures

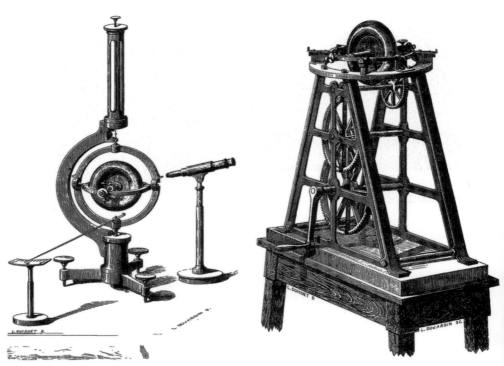

Figure 1: Drawing of Foucault's gyroscope (left), and of its launching device (right). Taken from the *Astronomie populaire* by Camille Flammarion.

Figure 2: Straight and Level.

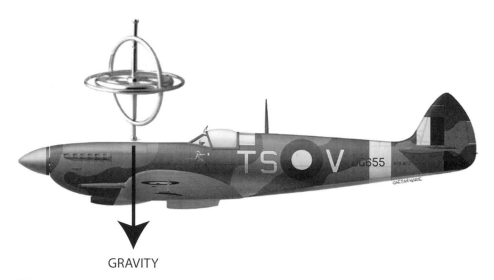

GRAVITY

Figure 3: Straight and Level.

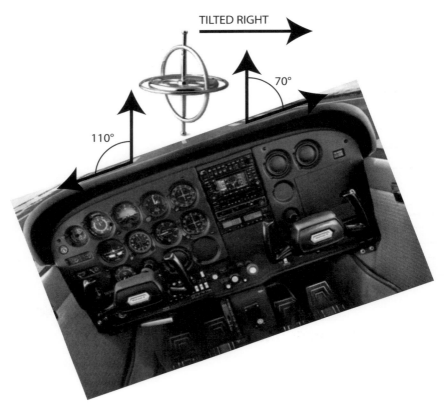

Figure 4: Left Turn, 20° Bank.

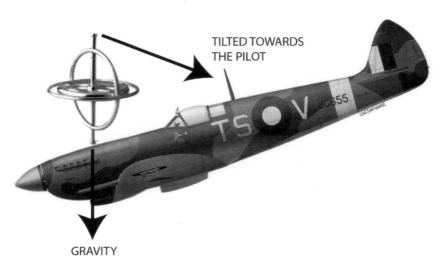

Figure 5: Diving.

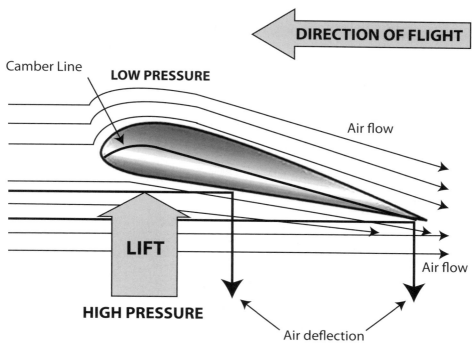

Figure 6: Creating Lift on a Wing.

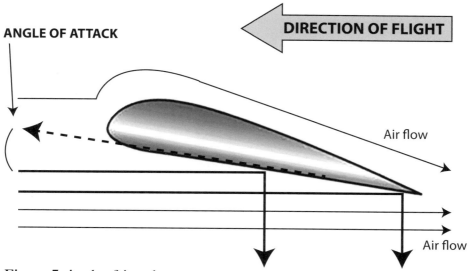

Figure 7: Angle of Attack.

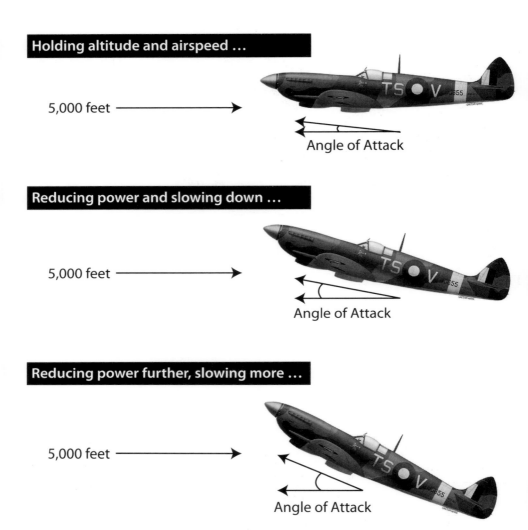

Figure 8: Slowing Down While Holding Altitude = Increased Angle of Attack.

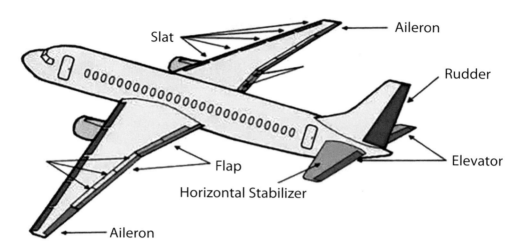

Slat

Aileron

Rudder

Flap

Horizontal Stabilizer

Elevator

Aileron

Figure 9: Control Surfaces.

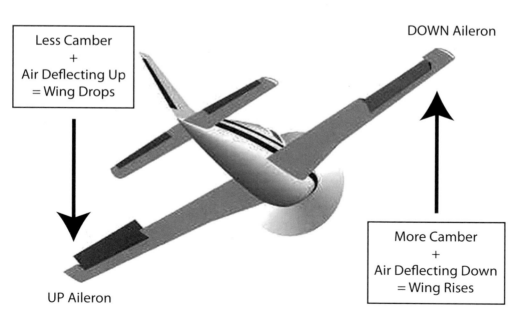

Less Camber
+
Air Deflecting Up
= Wing Drops

DOWN Aileron

UP Aileron

More Camber
+
Air Deflecting Down
= Wing Rises

Figure 10: Ailerons in Left Bank.

Figure 11: Tail Control Surfaces.

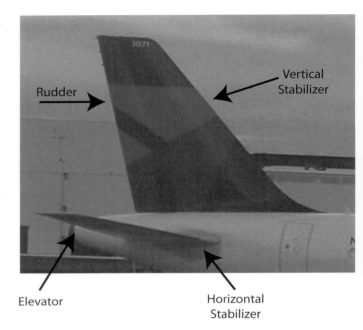

Figure 12: Horizontal Stabilizer for Trim.

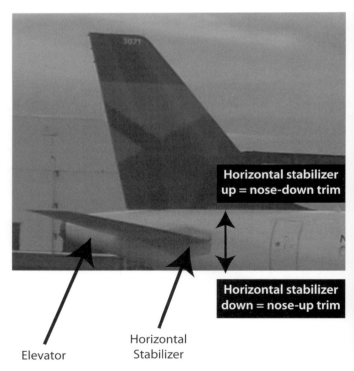

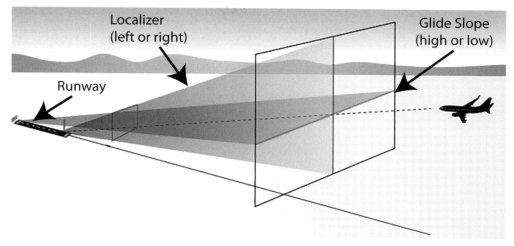

Figure 13: Instrument Landing System Radio Beams.

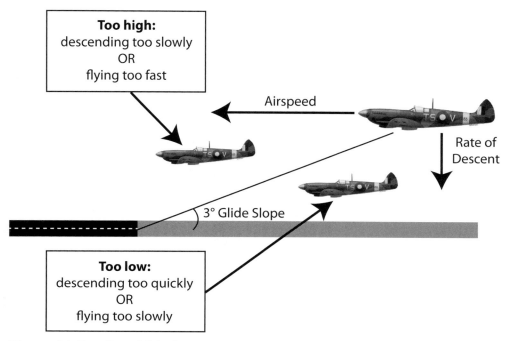

Figure 14: Landing Glide Slope.

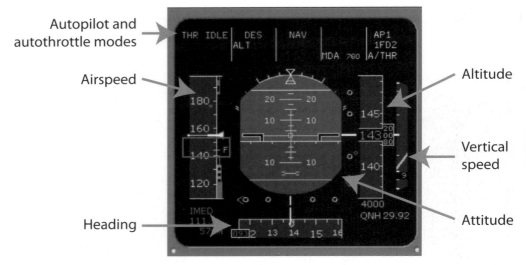

Figure 15: Primary Flight Display.

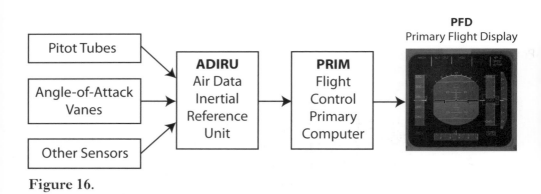

Figure 16.

Primary Flight
Display

Navigation
Display

Upper and Lower
ECAM displays

Land Gear

Throttle

Sidestick

Flaps

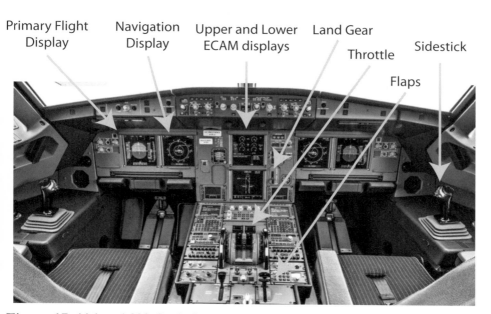

Figure 17: Airbus A320 Cockpit.

Mode Control
Panel (MCP)

Primary Flight
Display

Navigation
Display

Throttle

Land Gear

Flaps

Yoke

Stabilizer
Trim
Cutout

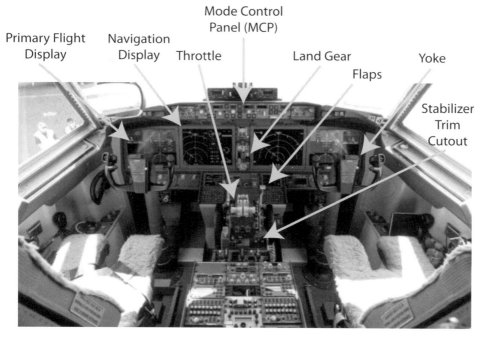

Figure 18: Boeing 737 MAX Cockpit.

Part I

Hands Up

Chapter 1

Cleared for Takeoff

On the last night of her life, 24-year-old First Officer Rebecca Shaw was fighting a head cold. It had hit her that morning and was causing her to sniffle, but it wasn't bad enough to keep her from her job as a copilot for Colgan Air.

That February evening she was paired with 47-year-old Captain Marvin Renslow in the cockpit of Colgan Air flight 3407, flying a Bombardier Q400 turboprop from Newark Liberty International Airport in New Jersey to Buffalo, New York. Behind them in the narrow cabin were forty-five passengers and two flight attendants. Already running two hours late, it is likely that few of them were happy.

At 8:30 pm they were finally given permission to taxi to Runway 22-Right.[1] High winds in the New York area were pummeling airline schedules and Newark's taxiways were jammed with planes. Renslow and Shaw chatted idly as they snaked slowly along the conga line. Outside their cramped cockpit the sky was clear and the air unseasonably warm. But while it was surprisingly nice in Newark, they expected light snow and freezing cold near Buffalo.

Forty-seven minutes later Colgan 3407 was close to the runway. Shaw switched her mike to the cabin public address system. 'Folks,' she said, 'it looks like we're number two for departure. Should be pretty quick here. Like to have the flight attendants please take their seats. Thank you.'

A controller in Newark's Air Traffic Control Tower radioed them. 'Colgan thirty four oh seven runway two-two right at Whiskey position and hold.'

Renslow steered the plane onto the runway from taxiway Whiskey, lined it up with the centerline, and then with his toes on the brakes went through the short Before Takeoff checklist with Shaw.

---

1 Runways are named for their approximate compass heading, with the zero omitted. Runway 22 is the Newark runway pointing to a heading of 220° (it is roughly 219° but was rounded up). Newark has two of these runways parallel to each other a few hundred yards apart, hence the designations 22-Left and 22-Right.

Just as they finished, the tower controller called them again. 'Colgan thirty four oh seven, runway two-two right at Whiskey. Winds three-zero-zero at one-niner. Cleared for takeoff.'

Shaw acknowledged immediately. 'Cleared for takeoff Colgan thirty four zero seven.'

Renslow quickly briefed Shaw. 'Alright cleared for takeoff. It's mine up to two thousand heading two-seven-zero after departure. Here we go.'

He would fly while Shaw would monitor the cockpit instruments and make the required call-outs during the takeoff.

Releasing the brakes, Renslow pushed the throttles up to takeoff power and the noise level rose in the cockpit. 'Check power,' he said as the plane accelerated down the runway.

Shaw glanced at the engine instruments. 'And power checked,' she said.

Four seconds later Shaw called out their airspeed. 'Eighty knots.'[2]

'Eighty,' Renslow confirmed.

A few more seconds, then Shaw said, 'V-One,' letting Renslow know they had passed the point where in an emergency they could stop on the runway. If anything went wrong now they were committed to flying.

One beat more and they reached flight speed. 'Rotate,' Shaw ordered.

Renslow *rotated* the plane, pulling back on the yoke and tipping the nose into the air. As the plane left the ground they could feel the nose wheel tires still spinning under them.

'Positive rate,' Shaw reported. Cockpit instruments showed they were climbing.

'Gear up,' Renslow responded.

Shaw lifted the landing gear handle, retracting the wheels into their wells. The gear doors closed with a satisfying thump.

The Newark Tower controller radioed them to make a slight right turn and handed them off to New York Departure Control. Checking in with New York Departure, Shaw received permission to climb to 10,000 feet.

As they powered skyward Renslow, unable to help himself, suddenly said, 'Wee, this is fun.'

'Yeah,' Shaw agreed.

Pilots love to fly.

Climbing through 8,000 feet, Renslow pressed a button on the instrument panel.

---

2 Globally, aviation uses *knots* for speed and *nautical miles* for distance. One knot equals one nautical mile per hour. One nautical mile is exactly 1.151 statute miles (statute miles are used for speed and distance in the US and UK). While on the subject, most of the aviation world uses *feet* for altitude.

'Autopilot's engaged,' he said.

'Alright,' Shaw said.

Slipping his hands onto his lap while the autopilot took Colgan 3407 to Buffalo, Renslow would not touch the yoke for the next 53 minutes and 55 seconds.

But when he did touch it again he would do the wrong thing, with disastrous consequences.

Chapter 2

Born to Fly

Humans have always wanted to fly. Many of us, anyway. And why not? Being able to go anywhere at will in three dimensions, like a bird, is absolutely thrilling. Pilots would be the first to tell you how incredible it is.

I am one, so I know the feeling. Like the vast majority of pilots, I have wanted to fly planes since my earliest memories. As a kid I looked up every time a plane flew overhead. I was raised in New York under the approach path to one of John F. Kennedy International Airport's runways, so I spent a lot of my youth looking up at planes in the sky.

Humankind's mastery of the air took most of recorded history to accomplish. Greek mythology has it that over three thousand years ago Daedalus fashioned wings for himself and his son Icarus to escape captivity in Crete. Though it worked for them for a while, it didn't end well when Icarus flew too close to the sun and became the world's first aviation fatality.

Nothing much happened in our quest to fly until the Industrial Age, when scientists, mathematicians, and inventors discovered the theories and created the tools that led to Wilbur and Orville Wright's first powered flight in 1903. Their Wright Flyer wasn't airborne very long, just 12 seconds, and it flew only 120 feet in a straight line. But it was a start.

It took another ten years before a pilot purposely flew a plane upside down. Then came the First World War and airplanes became weapons. After that, the things they could do in the sky finally began approaching the freedom birds have always enjoyed.

When pilots learn to fly, whether in the military or at the local airfield by their house, they begin with small single-engine propeller planes. They start in good weather, when it isn't very windy and they can see at least a few miles in any direction. Then training shifts to flying in clouds and bad weather. As they demonstrate proficiency they graduate first to twin engine planes, then to jets. Standards of excellence increase each step of the way. What is good enough for a lower-level license won't pass muster the next level up.

By the time pilots are qualified to command an airliner they are proficient, capable and safe in all flying conditions. And one thing has never changed – their passion for flying.

A small minority of commercial airline pilots earned their certificate for reasons other than the visceral love of flying. Maybe they read an ad for an airline pilot training program, triggering a sensation they had never felt before. Or perhaps they believed it could launch them into the upper middle class. Regardless of why they ended up in the cockpit, they didn't survive the gauntlet of flight tests and written exams unless they had proved to their instructors, examiners, and especially to themselves they could handle a plane.

Much has been written recently regarding the competence of pilots flying for small and poorly financed South American, Asian and African airlines. Pundits and industry observers claim they don't have sufficient training – that it is both not enough and not the right kind. They insist it is wrong that copilots outside the US can get a job with only a few hundred hours in the air, arguing it is far too little to acquire the skills needed to control a modern airliner.

All that may be true to some extent, though laying out both sides of the debate would take a few chapters, at least. And whether it is a bad thing is another question entirely. If their nation's aviation authority awarded them their seat in the cockpit and their airline put them on your flight, they are good enough to get you where you are going.

But 'good enough' is not what you want when you are wedged into economy bouncing in turbulence. You want Chuck Yeager or Jacqueline Cochran or Chesley Sullenberger at the controls.[3] But while you may want that quality of pilot – and your captain may actually be no less talented – you do not really need it. You are not engaging in air-to-air combat or breaking world speed records. You are just trying to get to Birmingham or Frankfurt or Disneyworld for a week's vacation with the kids. Besides, these days commercial airline pilots have a weapon at their disposal enabling them to fly the biggest airliners as well as the best old-school jet jockeys.

---

3 Chuck Yeager was an American fighter ace during the Second World War. On 14 October 1947 he became the first pilot to fly faster than the speed of sound. Jacqueline Cochran was an airplane racer in the 1930s, and head of the WASPs during the Second World War (Women Airforce Service Pilots – an all-female flying organization attached to the US Army Air Forces). On 18 May 1953 she became the first woman to break the sound barrier. Chesley Sullenberger III was a former US Air Force pilot and captain of US Airways flight 1549, which landed in New York's Hudson River on 15 January 2009. The flight is covered in Chapters 51 and 52.

That weapon is *automation*.

Automation is wonderful, a vital necessity we cannot live without. It is in our kitchen microwaves, in our cell phones, in the elevators we take and the cars we drive. More than ever, it is also in the planes we fly.

Aviation automation has progressed from the first rudimentary *automatic pilots* that kept the wings level with the ground, to the latest Flight Management Systems that take airliners from the foot of the departure runway to the end of the arrival runway. Commercial airliners today are ecosystems of specialized computers following programmed routing instructions and reading data from sensors throughout the airplane, all working to keep the plane safely on course, at the right speed and altitude for each leg of the route, for many hours at a time.

The pilots' primary job on these highly-automated jetliners is not to *hand-fly* the plane – that is, to control it with hands on the yoke and throttle, and feet on the rudder pedals. They are not mimicking Yeager, Cochran or Sullenberger. Instead, their main job is to monitor those computers, making an occasional minor adjustment here and there when needed. A pilot's real challenge in modern commercial aviation is pressing the right button at the right time.

And that is important: it has got to be the right button *every time*. No mistakes. A mis-pressed button or a mistyped instruction can upset that ecosystem. Same with a failed sensor, any one of which can play havoc with the plane's computer-controlled rhythms.

Unlike with a modern automated factory, when something goes wrong in the air the operator – the pilot – cannot stop the machinery and investigate. When a plane does something unexpected or a warning light blinks on at 37,000 feet, a pilot must instantly shift from monitor/button-pusher to trouble-shooter/problem-solver/flying-ace – and that assumes the pilot has even *noticed* the warning light.

Making that shift, as you will read, is surprisingly hard to do.

Automation has unquestionably made flying safer, though it is difficult to find irrefutable confirming statistics. More planes take to the air each day than ever, and aviation accident rates are lower than ever, so simple mathematics yields the lowest rate of accidents-per-flight in aviation history.

But should automation get all the credit? Or is the industry's safety record the result of better-trained pilots? Or sturdier aircraft?

Of the three, automation has made the greatest leaps over the years, so it deserves the bulk of the credit. But it is almost impossible to review a near-accident, a non-crash, and conclude that if automation had not been present, that plane *would have* crashed. We can hypothesize automation's role in averting the catastrophe, but we cannot credit it with absolute certainty.

The reverse, however, can be done. We have all read on occasion that '*automation* caused the plane to crash.' Whether it failed, did something unexpected, or did not do something it should have, automation is often the prime suspect.

Chapter 3

Human Factors

Aviation automation helps pilots in two ways. It keeps them extremely well-informed about the state of their plane and flight, and it makes controlling their plane easier and safer. Both functions are immensely useful, but both have their unsafe downsides. I'll take them one at a time.

In theory, automation keeping pilots informed improves their *situational awareness*, their overall knowledge of their plane and flight – where they are, where they are going, and the condition of their aircraft. It reduces distractions by monitoring the plane's systems with a degree of hypervigilance no human can match, and alerts cockpit crews when somthing has gone wrong, enabling pilots to focus on the important things, like navigation, other planes in the sky, en-route weather, and their passengers. Pilots are kept up to date in a concise and efficient way about everything they need to know.

But this function has a built-in flaw. When something goes wrong on a highly-automated airliner, it can inundate its cockpit crew with information. Most of what it is throwing at the pilots is right. But sometimes it is overwhelming, other times useless, and occasionally it is flat-out wrong. In the chapters ahead we will see where automation was unable to simply declare, 'XYZ is broken. Here is the solution to get you back on the ground safely.' Instead, the deciding factor in surviving was the cockpit crew's experience and coolness as they struggled to understand and deal with their predicament.

Then there is the opposite problem: information going dark, no cockpit warning message or instrument panel light announcing something had just gone terribly wrong. Pilots can become so accustomed to automation's hand-holding that when a system fails they are clueless how to solve it without clear and succinct notification by the plane's computers.

What if an aircraft manufacturer installs software it believes is making a plane safer, protecting the pilots and passengers, but it appears nowhere in the airplane's manuals? With no information to work with, if that software fails, how is a pilot supposed to trouble-shoot *that*?

That is exactly what happened to pilots of Boeing's state-of-the-art 737 MAX commercial airliner (Boeing capitalizes all three letters). Boeing didn't tell anyone about an entire software suite designed to iron out a handling problem caused by the plane's engines nacelles – not the powerplants themselves, but their housings. Boeing had placed professional pilots in a position of not knowing why their airplane was behaving in an unexpected and terrifying way. It was inexcusable.

It was not the first time Boeing had kept pilots in the dark. The company did not bother mentioning an important aspect of the automation on one of the MAX's predecessors, the 737-800, as well. That led to a crash in Amsterdam where people died. We will talk about these accidents in the chapters ahead.

As for making planes easier and safer to handle, an autopilot can make anyone fly perfectly. But there is a difference between a human at the controls and a machine. A pilot hears the whine or roar of propellers and jet engines, the muted howl of the wind through the windows and airframe, and feels a plane's movement about the sky (even though those feelings are not to be trusted – more on that when we talk about flying in clouds).

On the other hand, autopilots and computers only know what their sensors tell them. Sensors fail, sometimes completely, though other times they just lie, or disagree with other sensors. And even when sensors are working correctly, monitoring software can encounter a problem it cannot handle because its programmers didn't anticipate it.

Automation creates another problem: it takes flying out of the hands of pilots and puts it in the hands of the computer programmers who wrote the software dictating what the plane will do in every situation. Taking the plane out of a pilot's hands lowers a pilot's proficiency level and raises the risks on every flight.

Proficiency drops because of a simple and direct correlation: less time hand-flying means less time practicing the sport, the skill, of flying. In every endeavor requiring more than luck – baseball, football, chess and checkers, even gin rummy – if you don't play, your skills will decline.

It is no different for pilots. Multiple studies confirm it. Every aspect of a plane's flight requires perishable skills that can be lost if not practiced. For a pilot hand-flying, it is not easy keeping a plane flying a perfectly straight line while holding altitude. It takes concentration and effort, even on a clear day without a bump in the sky. It is more difficult to turn while holding altitude, or to climb or descend to a new altitude while turning.

Now combine those maneuvers with turbulence, clouds, and rain or snow. Add in an electrical failure, and then perhaps an engine fire. Oh, and now the pilot needs to land the plane.

Automation can handle every bit of that. But what happens when automation fails? If these situations are not hand-flown in simulators for practice regularly, passengers are left trusting their lives to a pilot who is not as sharp as they need. These days that is the case all too often. When sensors fail and software goes awry, I don't want my pilots out of practice. I want them having rehearsed that scenario countless times so that when things go to hell the maneuvers to fly us to safety are in their DNA, hardwired in their brains.

So far I have noted two of automation's dangers: that information it provides might be too much, too little, wrong, or non-existent, and that it reduces pilot proficiency.

Here is a third danger, no less serious than the first two. Aircrews in automated cockpits are going hours without turning a knob, touching a screen, or pressing a button. Leave humans in a state of observation for hours on end and their senses dull, their reflexes slow, their attention wanes. Again the risks go up as their situational awareness fades.

Commercial pilots' physical and psychological operating environment has become a major area of study and focus in the aviation community. Known as *human factors*, this is especially important now, with automation playing such an outsized role on every commercial airline flight. Automation's effectiveness is dependent not just on the engineers and programmers who design and build computers and software, but on the interconnected human factors in play while pilots spend almost all their time watching the electronics fly their plane.

Everything from the tactile feedback of the controls, to the size, shape and color of cockpit instruments and displays, to the sounds made by cockpit warnings, are being studied to learn their impact on air crews. We would like to believe pilots sense and react to everything happening around them, but we have learned – at the cost of lives – they don't. They miss things, sometimes very important things.

Getting human factors right is as important as using the right fuel.

The automated cockpit and the computer-controlled airplane are more dangerous than we expected. How did we get here? How did we arrive at a point where pilots are button pushers without – for some of them – the skills they once had?

It began over 150 years ago.

Chapter 4

Gyroscope

In the first years of powered flight planes had no cockpit instruments, nothing telling pilots how fast and how high and in what direction they were flying. They didn't even have a cockpit. Pilots sat in front, the air whipping their faces letting them know when their machine was rolling fast enough over the grass field to take to the sky. At forty miles an hour, typical flight speeds for the earliest aircraft, sitting up front with a good pair of goggles and a warm leather coat wasn't so bad.

But soon planes were flying faster and higher than leather, goggles and guesswork could handle. Pilots needed cockpits and planes needed instruments.

I am not overreaching by claiming that without the gyroscope, the history of the twentieth century would have looked very different. No ship, submarine, plane or spacecraft could have sailed, submerged, flown or navigated safely without it.

The first gyroscopes appeared in the early 1800s. They are direct descendants of the spinning top, the child's toy that has been around for over five thousand years. Inventors and scientists – and children – knew that until its spinning slowed, a top would remain upright and motionless as it twirled. At its core the gyroscope is that same spinning top, except surrounded by a thin metal frame. But while they were fascinating to kids, no adult could identify a real use for them.

That changed in the 1850s when French physicist Léon Foucault was searching for a foolproof way to demonstrate to the masses that the earth rotated. The Paris-born Foucault had originally trained as a doctor, but an intense fear of blood unavoidably forced his educational pursuits to be directed elsewhere. He became a prolific inventor and is best known today as the creator of the *Foucault Pendulum*, something we are all taught about in middle-school science class.

This might sound familiar from back then. In 1851 Foucault hung a 62 lb brass ball under a 220-foot wire and set it swinging in a long sweeping arc.

It was so heavy and its swing so pure that as it silently swished back and forth in place, the earth turned under it. If you stood directly in front of it as it swung, first making certain the brass ball didn't clock you, within a few minutes it would no longer be swinging at you. You, and the earth under you, had moved while the swinging ball remained in place. Its position in space didn't change – *yours* did. You had witnessed the earth rotating.

Not satisfied with one proof the earth rotates, in 1852 Foucault saw in the gyroscope the possibility of a second. While working with it he not only invented the modern version of the gyroscope, he coined its name, which roughly translated from Greek means *see rotation*.

Many of us remember gyroscopes, how if you wound the string that came with it around its axle and then pulled with all your might, the center wheel would spin furiously and the thing would balance perfectly on one of its thin feet, standing as erect as a toy soldier.

Place the spinning gyroscope on top of a book you held parallel with the floor, and it would dutifully stay where you put it, spinning merrily, its axle pointing directly at the ceiling. Now here is the interesting part: if you moved the book, tilting it this way and that, the gyroscope would remain pointing at the ceiling. No matter how you angled the book in your hand, the gyroscope's axle would continue pointing straight up.

Thinking big, Foucault realized a spinning gyroscope was not really pointing at the ceiling. Instead, it was actually utterly immobile in three dimensions. It remained precisely where it was placed, continuing to point at whatever it was pointing to, regardless of what the world around it was doing. Put differently, just like his pendulum *swinging in space*, a gyroscope was *maintaining its position in space*.[4]

Foucault presumed if he suspended his gyroscope properly – meaning with no friction or as close to 'no friction' as he could come in 1852, with nothing to stop it from maintaining its place in space – as the earth rotated, he could show that rotation against the immobile gyroscope.

Practical electric motors didn't exist in 1852, so Foucault built a crank-and-gears contraption that got his gyroscope spinning at a remarkable 12,000 revolutions per minute. [FIGURE 1] Once spinning, he focused a microscope on a point on the gyroscope and, as he expected, soon saw movement. The microscope was drifting away from the gyroscope.

---

4 Engineers will know I am leaving out a massive amount of detail about the gyroscope, for instance precession and angular momentum, to name just two topics. Though important to understanding a gyroscope's behavior, they are beyond what we need to know to understand its use in aircraft automation.

Exactly like with the pendulum, while the gyroscope held steady everything else moved: Foucault, his microscope, the table it sat on and the earth beneath all of them had shifted together. As the name said, he was 'seeing rotation.'

Why do we care? Because a gyroscope's stability in space means if you point it at 'up' it will keep pointing at 'up' while you and the vehicle you are flying in turn, climb, roll, descend, or loop. Go into outer space and point it at the earth, and while your spaceship swings around the back side of the moon you will always know how to find home. Foucault's pendulum was also stable in space, but it would not fit inside a space capsule or a plane cockpit, and it needed gravity to work.

Very few of us will ever get to outer space, but we can hop into a small plane to watch a gyroscope in action. Start the engine, take off, and climb to a safe altitude. Start your gyroscope spinning and place it upright on the glareshield (in a car that would be the top of the dashboard). It will stay there spinning and pointing straight up. [FIGURES 2, 3]

If you bank the plane the gyroscope will appear to you, sitting in the pilot's seat, to be leaning the opposite way. Bank the plane left and the gyroscope will look like it is leaning right. Banking right will make the gyroscope look like it is leaning left. That is because while the plane has tilted, the gyroscope continues pointing straight up. [FIGURE 4]

If you push your plane into a dive, the gyroscope will look like it is leaning towards you. Same reason as before: you've tilted earthward, not the gyroscope. If you climb, it will appear to be leaning away from you. [FIGURE 5]

Pilots don't look at gyroscopes while flying. They look at instruments *connected* to gyroscopes. But first someone had to invent those instruments.

That is where the father and son team of Elmer and Lawrence Sperry come in.

Chapter 5

Elmer Sperry

Elmer Ambrose Sperry was born in Cincinnatus, New York, on 12 October 1860 to Stephen and Mary. He was their oldest and, it would turn out, their only child, as Mary died the day after Elmer was born, of complications from childbirth.

Elmer was a direct descendant of Richard Sperry, an Englishman who had come to America in 1634. Richard had earned his living as a farmer not far away, near New Haven, Connecticut, but he had made his mark as one of the men who helped hide two of the Regicide Judges, famous for condemning British Sovereign Charles I to death.[5]

Raised in Cortland, NY, Elmer's talent for invention became obvious at a young age. In 1879 he entered nearby Cornell University, where he worked in a lab designing electric dynamos, state-of-the-art in power generators in those years. From dynamos he moved on to arc lamps, the bright lights of the day powered by dynamos.[6]

Then in 1880 Elmer left Cornell to strike out on his own. Not yet 20 years old, he moved to Chicago to launch a company manufacturing and installing what he knew best – arc lamps and dynamos. Within a few years his business grew large enough to attract the notice of the founders of what eventually became Commonwealth Edison. He sold to them and turned his attention elsewhere.

Elmer married Zula Augusta Goodman, the daughter of a local Chicago preacher, in 1887. On their honeymoon he met a mine operator. They got

---

5 The Regicide Judges were 59 men who had signed the death warrant of British king Charles I in 1649 during the English Civil War. The warrant was carried out, but when the monarchy was restored in 1660 the death warrant signatories found themselves under their own death warrants. Three escaped to the New World. In the spring of 1661 two of them hid for nearly a month in a cave adjacent to land farmed by Richard Sperry.

6 Arc lamps produce light from an electric arc jumping between two electrodes. Used in the first practical electric light bulbs, the technology was superseded by *incandescent* light bulbs, that use a filament in place of the arc.

to talking, and Elmer saw his next opportunity. When the Sperrys returned to Chicago, he put himself in the mine equipment business by inventing, and then manufacturing, powerful electric mining machinery. After that venture, came trolley cars, locomotives, and even an electric automobile.

In 1914 one of Elmer's companies invented an arc lamp for the US Navy ten times brighter than those then in use, enabling massively powerful seaborne searchlights. In an unexpected byproduct of his invention, cameramen in the nascent motion picture business used his lamps to light sound stages, recreating daylight indoors. Elmer then miniaturized the light technology to produce bulbs for movie theater film projectors, yielding brighter, clearer images on screens.

In all, Elmer Sperry held nearly four hundred patents and founded eight companies. Lighting cities, mining coal, helping the Navy, and playing a role in Hollywood made him wealthy. But stormy weather on a transatlantic passenger-ship voyage in 1898 got him working on a device that would one day lead directly to inventing the autopilot for airplanes – after, of course, Wilbur and Orville invented the airplane.

Elmer had been so uncomfortable as his ocean liner tossed around in heavy seas that he set about seeking a device to stop – or at least dramatically reduce – a ship's pitching (up and down movement of the bow and stern) and rolling (side to side motion) in stormy waters. His engineering mind intuitively figured out that gyroscopes should play a role. He had become an expert in them while inventing the *gyrocompass*, a gyroscope-based compass that could find North without being affected by the metal construction of a ship. To manufacture his gyrocompass he had founded the Sperry Gyroscope Company in 1910. So Elmer Sperry, more than most, understood their behaviors and properties.[7]

Eventually Elmer found his answer to a ship's pitching and rolling. Called a gyroscopic stabilizer, or *gyrostabilizer*, his invention was a pair of massive 25-ton gyroscopes nine feet in diameter strategically placed deep in the bowels of a ship. When sensors attached to the spinning gyroscopes

---

7 A gyrocompass could find, and hold, North without using earth's magnetic field, as magnetic compasses do, and so all-metal ships had no effect on its accuracy. Sperry invented his gyrocompass in 1908, but two years earlier Hermann Anschütz-Kaempfe, a German inventor and businessman, invented one with slightly different technology but the same ability to find and hold North. Anschütz-Kaempfe sued Sperry for patent infringement. In their trial in Germany, Anschütz-Kaempfe called Albert Einstein as one of his key witnesses. Not surprisingly, he won the case. But they were also engaged in simultaneous lawsuits in the US and UK, and Sperry won those. Plus, the US was about to go to war with Germany, so in the end the suits had no impact on Sperry's business.

detected the vessel starting to roll or pitch, they would instantly shift to counteract the movement by their sheer weight and the momentum of their spin. Though it worked, the device was incredibly heavy – as much as five percent of the weight of the ship – and broke down often. Nautical designers soon discovered that a better way to keep ships stable was placing fins along their sides, but by then Elmer had gained much useful knowledge about gyroscopes.

What does this have to do with aircraft autopilots?

As with ships, an airplane's *pitching* is the up and down motion of its nose and tail, and *rolling*, also called *banking*, is tilting from side to side. But unlike with ships, pitching and banking/rolling in airplanes is completely necessary. Planes *pitch* up and down to climb or descend, and *bank* to turn. But those movements are unwanted while flying in a straight line.

Almost no one saw any value in a piece of machinery that would keep a plane on that straight line while the pilot took his or her hands off the yoke during a long flight to rest, or perhaps to focus on a map.

No one except Elmer Sperry's third child, Lawrence.

Chapter 6

Lawrence Sperry

Elmer and Zula Sperry had four children – Helen, Edward, Lawrence, and Elmer Jr. While all four eventually worked in the family business, Lawrence seemed to have gotten both inventor genes from his father and risk-taking genes from his ancestor Richard, the seventeenth century farmer.

Lawrence was born in Chicago on 22 December 1892, before the family moved to Brooklyn, New York. He was so keen to fly as a teenager that with help from his kid brother Elmer Jr he built a glider in the basement in their vacation home in Bellport, Long Island. When they had completed construction they discovered the plane was too large to slip out of the front door, so they removed the family home's big bay window. After the plane exited the home the brothers didn't quite get the window back in place.

Elmer Sr could see where Lawrence was heading, so after making his sons pay for the damage he got Lawrence a job with Glenn Curtiss, an early aviation pioneer. Curtiss was building planes and teaching flying in Hammondsport, New York, near Cortland, where Elmer had grown up. In 1913 Lawrence was awarded pilot license #11 from the Aero Club of America.

Handsome and brilliant, five years later he would marry silent film star Winifred Allen. But before settling down to married life he earned credit for one of aviation's most important advances, and also for one of its most notorious: the autopilot in 1914, and the Mile High Club in 1916. Though the autopilot came first, and it plays a leading role in the Mile High Club story, the latter tale is easier to relate.

On 22 November 1916, Lawrence was flying off the south shore of Long Island, New York, in his autopilot-equipped Curtiss Flying Boat biplane. His passenger was 27-year-old Dorothy Rice Peirce, who had received her pilot's license a few months earlier. Her husband, the well-known artist Waldo Peirce, was away in France driving ambulances to play his part in the Great War. Either because Lawrence was demonstrating his autopilot, or because he was demonstrating something else entirely, the plane crashed into Great South Bay.

Duck hunters found the couple wearing not a stitch of clothing. Lawrence explained with a straight face that in the process of making a normal landing on the water, the plane hit a stake planted into the bay's shallow bottom with such force that the subsequent crash had torn off all their clothes. Given his minor injuries, his story is a bit hard to believe.

Dorothy, meanwhile, had broken her pelvis. In her version of the accident, she thought Lawrence was doing aerobatics of some sort – 'stunts,' in her words. Relating the story to a reporter, she said, 'By the time I realized the truth, we were under the water.' She went on to explain she subsequently learned one of the plane's 'control wires' had broken, causing the crash.

No matter how they ended up naked in the water, and though they were reportedly flying at only 600 feet – not close to a mile high – historians who care about these things took the leap and credited Lawrence and Dorothy as founding members of that unique club.[8]

Now on to the autopilot.

Lawrence intuitively recognized the value of an on-board device that could keep a plane flying in a straight line, not deviating left or right, up or down, without a pilot touching the controls. It made sense to him. Using pilot vernacular, Lawrence sought to invent a device that would keep a plane flying *straight-and-level* without pilot intervention.

Straight-and-level sounds easy, but it is not at all: it is the first true test of piloting skill. Putting *level* in perspective, it doesn't seem like a big deal if an airliner flying at 35,000 feet drifts up a little to 35,050 feet or slips down a bit to 34,950 feet. But each of those 50 foot altitude changes is the height of a five story building. That is way too much error for a professional pilot.

Flying perfectly *straight* is equally important. Navigation's one-in-sixty rule-of-thumb states if a plane is just one degree off course, then over a sixty mile flight it will be one mile away, to the left or right, from where it should have been (the actual error would be 1.05 miles, but the rule's estimate is close enough). Over a 200 mile trip – the distance between New York and Boston, or London and Paris – that one degree of error will take a plane more than three miles off course. That is unacceptable.

A pilot in a jet fighter flying straight-and-level in formation with another jet has to stay within three feet of his partner. The tolerance is eighteen inches for pilots in aerobatics teams like the US Air Force's Thunderbirds or the US Navy's Blue Angels. A pilot who can't fly straight and level will never be able to do that.

---

8 Dorothy made a full recovery, then divorced Waldo and soon after married Hal Sims, a ranked bridge player.

LAWRENCE SPERRY

Many years ago my uncle flew F-106 Delta Dart supersonic interceptors for the US Air Force. Long after he had retired he was over my house and I was showing him a new flight simulator video game I had just bought. I set the simulator to mimic a P-51 Second World War fighter plane. Nimbly holding the toy joystick in his right hand, the first thing my uncle did was try keeping the simulated airplane straight-and-level. No turns, no gaining or losing altitude, and definitely no fancy dog-fighting maneuvers. His training taught him if he could master straight-and-level, he could master the plane. Eventually he got it, and then he added turns, and climbs and descents, building on his experience until he felt he could properly control the computer-based machine.

To meet the challenge of straight-and-level flight, Lawrence reasoned that if a gyroscope could sense and stop the pitch and roll of a ship, it should be able to sense and stop the pitch and bank of an airplane. But he also reasoned that the mechanism would need to operate differently on a plane than aboard a ship. A huge and heavy hunk of spinning metal in the belly of a plane wasn't going to cut it.

Lawrence returned to Foucault's first use for a gyroscope in 1852: it stayed frozen in space while the table it sat on turned with the rotation of the earth. If the gyroscope was truly rock-steady in space, it could be placed in an airplane with sensors rigged up to record the plane's movement. The plane would pitch and bank, the gyroscope would remain in place, and sensors would detect the difference.

Confident he was heading in the right direction, Lawrence designed an apparatus using four gyroscopes. Two monitored horizontal movement, two watched the vertical. He used two for each axis so they'd spin opposite each other, cancelling out their inclination to *precess*, their inherent tendency to wander away from the direction they're pointed. The gyroscopes were electrically spun at 12,000 revolutions per minute (coincidentally the same speed as Foucault's hand-cranked gyroscope) and mounted within two free-revolving rings, one horizontal, the other vertical. The ring pair gave the gyroscope quartet complete freedom of movement in three dimensions. Lastly, the gyroscopes were attached to sensors. Electricity came from a dynamo – Elmer Sperry's original specialty – with a back-up battery in the unlikely event the dynamo failed.

Once turned on by the pilot, if the gyroscopes' sensors recorded the slightest banking or pitching by the plane they would instantly trigger small servos, motors attached to the plane's *control surfaces* – the catch-all term for the ailerons, elevator and rudder, which we will discuss soon – moving the right surface by just enough to get the plane back to straight and level flight.

21

THE DANGERS OF AUTOMATION IN AIRLINERS

In addition to the four gyroscopes, Lawrence added one more sensor, an anemometer – a wind speed indicator – to measure the airplane's speed. Today it seems obvious that pilots should know how fast they are flying, but remarkably even in 1914 that wasn't the case. They could feel the speeding air brushing their faces and pulling at their clothes, and they figured that was sufficient.

Sperry's anemometer would not only tell pilots how fast their plane was moving through the air, it had two other important functions. First, it would ensure the plane flew fast enough to remain airborne. Second, it would fine-tune the servos attached to the control surfaces so they would not overreact when the gyroscopes' sensors signaled adjustments were necessary. Overreaction was, and is, dangerous – as much today as in Sperry's time. Just like with a car's steering, it takes less control surface movement to make a plane climb or turn when it is flying fast than when it is flying slowly. Without that fine-tuning, a plane going fast would over-respond and one going slowly might under-respond. Neither was safe. So Sperry installed the anemometer.

The entire package, including the gyroscopes, servos, anemometer, dynamo, and battery weighed 45 pounds, light enough to fit in the Curtiss Flying Boat Lawrence flew in the demonstrations.

Lawrence referred to his invention by the same name his father used for the ship pitch-and-roll solution: a *gyrostabilizer*. In actuality, he had invented the first *autopilot*, a name not used until decades later.

Lawrence began testing his gyrostabilizer in 1913 with help from the US Navy, whose ships were using his father's gyrostabilizers, gyrocompasses and searchlights (as well as other Elmer Sperry inventions). Soon he had it running well enough to show the world. The perfect venue was the *Concours de la Sécurité Aviation en Aéroplanes* (the *Airplane Safety Contest*), being held outside Paris, France, from January through June 1914. During those six months planes gathered at Buc aerodrome (now Toussus-le-Noble airport) twelve miles south-west of Paris, and gave demonstration flights past the judges over the Seine River by the town of Bezons thirteen miles to the north.

On 18 June 1914, 21-year-old Lawrence Sperry took off in his Curtiss flying boat to demonstrate his gyrostabilizer with his French mechanic, Emil Cachin, sitting in the seat to his right in the open cockpit. His parents Elmer and Zula had come to Paris to watch and were in the crowd by the river. In a series of fly-bys past the judges and his parents, Lawrence first stood up in the cockpit while holding his hands high in the air to show he wasn't touching the controls. The plane flew straight and true.

Then he circled around and flew past a second time, again standing with his hands high, but this time Cachin stepped out of the cockpit to stand six feet away on the right wing. Unbalanced on the right side, the plane banked and turned that way. Embarrassed, Lawrence sat back down, took the controls, made a few adjustments and swung back around to fly past the judges and his parents again. This time, in spite of Cachin's destabilizing weight on the right wing the Curtiss flew straight and level as Lawrence held his hands high. During the next pass Cachin crawled a few feet back towards the tail. The Curtiss flew steadily on, the gyrostabilizer perfectly adjusting for Cachin's presence in the back.

On the final demonstration pass Sperry reduced engine power and the craft slowed precipitously, which was detected by the anemometer. The gyrostabilizer pitched the nose down until enough airspeed had been gained in the dive to restore safe flight. Then it eased the nose back up to return the Curtiss to straight and level.

Lawrence's flight so captured the judges' imagination that a number of them asked to come aboard for personal demonstration flights. At the conclusion of the competition Lawrence was awarded first prize, 50,000 French Francs, just under $10,000 in that year's currency, beating fifty-six other entrants.[9]

Aviation automation had taken its first giant leap.

---

9 Adjusted for inflation, this would be approximately $250,000 in today's dollars. But if a competition like this were being held today, first prize would probably be $5 – $10 million, or even more.

Chapter 7

Lost Decades

Lawrence Sperry wasn't done inventing aviation gear after his flight before the judges and his parents in Bezons. Not quite as prolific as his father, Lawrence eventually put his name on twenty-three patents, including a few for some of the most crucial instruments in a cockpit, instruments still in use today. His contributions were cut short in 1923 when he was killed flying a plane of his own design across the English Channel. He was spotted three miles from shore, crash-landing into the cold waters after his engine had quit. When rescuers arrived they found his plane, afloat by the wings, but no pilot. His body turned up one month later.

Surprisingly, between 1914 and the 1940s no great technological improvements were added to the gyrostabilizer Lawrence had introduced to the world. Over those years engineers made gyroscopes spin faster and servos react more quickly, but the gyrostabilizer's *functionality*, what it could actually do, hardly changed.

Lawrence had misread the market. At the time, helping a pilot do his or her job held no appeal to aviation designers.

The aviation community firmly believed the best way to keep a plane safe in the air was to strap a competent pilot into the cockpit and keep distractions to a minimum. The *automatic pilot*, as Sperry's gyrostabilizer was soon being called, was considered one of those distractions. It was an interesting novelty, but real pilots did not need the help. If a pilot needed a moment to study a map, that is what multi-tasking was all about. So as the years went on, planes flew higher, faster, further, and with more passengers, but with humans at the controls one hundred percent of the time.

The one place aviation saw some value in the automatic pilot was in what was then called *blind flying*, what today we more frequently call *flying blind*. Like it sounds, it is when a pilot cannot see out of the cockpit window and is blind to his surroundings, for example when inside clouds. But it is more than that. It is any time the horizon is not visible, for instance on a moonless night high over a featureless desert, or in thick haze and smog. In these conditions pilots can develop *spatial disorientation.* They may not feel disoriented, but without a visible horizon they cannot accurately judge their plane's position in space. They have no idea if they are straight-and-level, pointing at the stars, or heading for the ground. In short, they can't tell which way is up. Spatial disorientation has killed thousands of pilots, and most of the time they never knew they were in trouble until it was too late.

Spatial orientation – the opposite of spatial disorientation – starts with your eyes seeing the horizon. At the same time, your inner ear's vestibular system and your body's proprioceptors (sensory points within muscles and joints that feel movement, weight and position) work in unison to detect motion, rotation and acceleration, and to maintain your balance.

Earth's gravity is a form of acceleration you are accustomed to feeling all the time. It is why you have weight. While sitting in your living room, you are not floating above your sofa because your body is experiencing acceleration. The acceleration is not much, exactly one times the force of gravity, abbreviated as *1g.* If you have ever stood in an elevator that suddenly shot upward, making you feel heavier than your normal weight for an instant, you were sensing acceleration of *more than* 1g. If you drag-raced the driver next to you at a red light, like I did as a kid on Long Island with my shiny new license, being pushed back into your seat after stomping on the gas was also a feeling of more than 1g.

You have also probably experienced less than 1g, a feeling of weighing less than normal. If your elevator suddenly descended and you felt as if you had left your stomach a few floors above, you know how less-than-1g feels.

A pilot who cannot see the horizon can't translate the g-forces he is sensing into an awareness of what is happening to him. He has no way of knowing if, for instance, a sudden feeling of 2g – twice the acceleration of gravity – is from banking steeply in a *turn,* or from pulling up steeply into a *climb.* They feel exactly the same. If the eyes don't *see* the bank or don't *see* the nose pitching up against the horizon, it is impossible to know what's causing the feeling.

Not convinced? A video showing a well-known aerobatics maneuver might change your mind. A barrel roll takes a plane from right-side-up to upside-down and back to right-side-up as it traces a corkscrew-shaped path in the sky. It is called a *1g maneuver* because the g-forces on the pilot throughout most of the maneuver are 1g, the same as sitting on your living room sofa.

In the video, Bob Hoover, a famous airshow pilot, places a glass of iced tea on the glareshield of his plane and then performs a barrel roll. The glass doesn't budge and the tea doesn't stir. Then he does another barrel roll, this time while pouring iced tea from a pitcher into his glass. He doesn't spill a drop. If the tea can't tell what just happened, with your eyes closed you can't either. You can find it on the internet.[10]

In aviation's early days, pilots learned the hard way that they needed instruments to tell them what their plane was doing in clouds, because their body could not be trusted to let them know. Yet when those instruments were invented – many by the Sperry family – pilots either could not, or would not, master the skills needed to read and interpret them. A 1920s-technology automatic pilot was the perfect solution. It would keep a plane properly oriented with the ground. But a plane's human pilot needed to trust it implicitly, and it needed to be utterly failure-proof, both tall orders in the '20s. So other than airmail pilots – *neither snow nor rain nor heat nor gloom of night* – aviators of the day usually stayed away from clouds and dark nights.

In 1930 Elmer Sperry died of complications from gallstone surgery. But his passing had little impact on aviation automation development, as two years earlier he had sold the Sperry Gyroscope Company to North American Aviation. He was out of the game.

The next year, Sperry Gyroscope finally produced the world's first commercially available automatic pilot (it wasn't until the late 1930s that the name was shortened, first to the hyphenated *auto-pilot*, and then eventually to just *autopilot*). Called the *Sperry A1 Gyropilot*, Eastern Air Transport, forerunner of the now-defunct Eastern Airlines, mounted them in its Curtiss Condor passenger planes. The next year, United Air Lines and Trans World Airlines put slightly upgraded *Sperry A2 Gyropilot* automatic pilots into passenger planes in their fleets.

The general public got a good look at the promise and potential of the autopilot in 1933 when Wiley Post, a famed aviator of the day, flew solo around the world. It took him nearly eight days, a record for the time. His flight was a huge deal, rating front page headlines in national newspapers. He credited his plane's Sperry A2 Gyropilot, which he had named 'Mechanical Mike', for lessening his workload and letting him sneak in a few minutes of airborne nap time. While it is not clear why his Sperry A2 needed a nickname, his single-engine Lockheed Vega airplane had a name, as planes

---

10 To be complete, though it is mostly a 1g maneuver, the g-forces on the pilot at the start and end are closer to 2g. A video example can be found at: youtube.com/watch?v=V9pvG_ZSnCc

and ships did and still do. His *Winnie Mae* can now be found in the Smithsonian's National Air and Space Museum in Washington DC.

In the Second World War, autopilots were installed on big British and American four-engine bombers, as well as on nimble single-engine fighters, to ease their pilots' physical efforts on long-range missions. On the bombers they were also rigged up to bombsights. Peering through bombing optics during the final seconds of a run to the target, a bombardier in the nose manipulated his plane's control surfaces through wiring connected to the autopilot, commanding it to turn the plane slightly left or right and refining the path to the ideal bomb release point.

For the bombers, fighters and commercial planes of the early 1940s, the autopilot made sure the plane flew straight and level, or turned a little left or right when asked. And that is all it did. That was only slightly better than what Lawrence Sperry's gyrostabilizer could do in 1914.

Chapter 8

Progress

The Jet Age began as the Second World War was ending. The new engine technology could propel planes at speeds and to heights propeller-driven plane designers could not imagine in the 1930s, when autopilots were so uninteresting to them. With the higher speeds and altitudes came higher pilot workloads. They had to navigate faster. Small problems spiraled out of control more quickly. The consequences of mistakes were more deadly. Both propeller and jet powered commercial airliners could now fly above the weather, but first they had to fly safely *through* the weather to get up that high, and then fly safely back through it again to get down.

Autopilots were the answer.

Manufacturers, now including not just Sperry Gyroscope, but Bendix and Minnesota-Honeywell (the predecessor of Honeywell International), rapidly improved the capabilities of their autopilots, and this time the aviation community welcomed it.

To start with, it became clear that while it was nice that the autopilot could keep a plane flying straight, it would be much better if it could keep a plane on a definite *compass heading*. And while remaining level was useful, holding a specific *altitude* was much more useful.

It may not be obvious there's a difference, but there is. If turbulence momentarily upset a plane flying straight and level on autopilot, after the bumps were done it would once again be straight and level. But would the plane be pointed at the same compass heading as before? Not necessarily, because an autopilot had no internal mechanism to lock in that heading. And though after a series of bumps it would resume flying level, would it be at the same altitude as before? Again maybe not. The turbulence may have dropped it or propelled it upwards by hundreds of feet.

It was a short leap by engineers to realize that if a pilot could see from the plane's compass that a course correction was needed, a sensor could detect the same thing. If a pilot could see that altitude was off, or airspeed was too fast or slow, sensors could as well. And so instruments showing heading, altitude and airspeed were rigged up with sensors connected to automatic pilots which could keep their plane on its course, altitude and speed. And since a bombardier in 1944 could adjust his plane's heading using the autopilot, it didn't take much thought to figure out that a pilot should be able to give an autopilot a new heading to fly by twisting a knob, and the device would turn the plane to the new heading and then hold it like it was on rails.

With engineers opening up to the expanding capabilities of autopilots, and pilots and the public finally accepting their help, progress in aviation automation advanced quickly. But the progress can't be appreciated if you don't know how an airplane moves about the sky. It is hard to see how automation might control a plane better than humans without at least a bit of that background. We will also need to know this when we look into the details of why crashes happened.

So we are going to take a short detour from the story.

Chapter 9

Wings

It doesn't require much special knowledge to gain a basic understanding of how planes fly. Our detour will take only a couple of chapters. A few diagrams will help. When we are done, we will move on to the role automation played in fatal plane crashes.

The first thing to know is, a plane in the air is a balancing act between two powerful forces, one well-known that we encounter every day, the other specific to aircraft.

The familiar force is *gravity*, the 1g force keeping you from floating above your sofa. That same *gravity* tugs at a plane, trying to drag it earthbound.

The other force is *lift*, pushing up.

Lift is manufactured by a plane's movement through air, and comes from two sources. The first is a wing's shape. A wing's top and bottom surfaces are curved, with the difference between the two surfaces called a *camber line*. Because the top surface is usually more curved than the bottom (the bottom is often nearly flat), the camber line itself is curved upward, and just to be slightly confusing, the curve is referred to as camber. [FIGURE 6]

For a plane in flight, the upward curve – the camber – causes the air above its wings to move faster than the air underneath. In the 1700s, Daniel Bernoulli, a Swiss physicist and mathematician, discovered a basic truth about water and gasses (like air): the faster it moves, the lower its pressure. For an airplane, that means the faster-moving air above its wings has lower pressure than the slower-moving air beneath them. The higher pressure underneath pushes upward towards the lower pressure. That upward push is *lift*.

Lift's second, and equally important, source is a wing's upward tilt into the airstream.[11] That tilt has a name: *angle-of-attack*, literally the angle at

---

11 For years the reason you're about to read was ignored completely, or its importance downplayed. That is because an older, dominant reason was used to explain why camber causes air above the wing to move faster than air below it: because air molecules have further to travel above the wing than below it. That reason has been completely debunked – it isn't true.

which a wing meets the airflow (or more literally, the angle at which the wing *attacks* the air).

When a moving wing is tilted up, air smacking into its underside is deflected downward. It is just like when you stick your hand out a car window at highway speed. Hold your hand out, flat but tilted up slightly, with a *positive angle-of-attack* as a pilot would say. If you use muscle power to hold it in place (at sixty miles an hour it is not easy keeping your hand perfectly still out the window), the wall of air hitting your palm will deflect downward. That wall of air is the same wall of air hitting the underside of an up-tilted wing. [FIGURE 7]

Like so many things in the physical world, what happens to that wall of air is best explained by Sir Isaac Newton. Newton's long résumé includes astronomer, physicist and mathematician, but he is perhaps best remembered for his Three Laws of Motion. First published in 1687, they describe in simple terms how objects behave, from feathers to apples to bowling balls.

His Third Law states: *every action has an equal and opposite reaction.* The air your palm is deflecting downward generates an equal-and-opposite force upward. Same with a wing: the air a wing is deflecting downward generates an equal force upward. That upward force is the second source of *lift*.

For a plane flying straight and level, the force of *gravity* pulling down exactly matches the force of *lift* pushing up. For a plane to climb or descend, that relationship has to change.

To descend, a pilot changes that relationship by reducing power. You may have noticed that on your last plane flight. Your pilot might have announced, 'We're beginning our descent into the Cincinnati airport,' and then the engine noise lessened. When power was reduced, the plane slowed down, air flowing above and below its wings slowed and air hitting the underside lessened. With both less pressure difference from the wing's camber and less downward deflected air, *lift* decreased. When that happened, *gravity* overpowered *lift* and the plane headed lower.

If your pilot had added power, the opposite would have happened. The speed of air flowing around the wing would have increased, so *lift* would have increased. The increased *lift* would have overpowered *gravity* and the plane would have risen.

There is also a way to increase lift without touching the power settings: by increasing the angle-of-attack. The effect of angle-of-attack on a plane's flight is *crucial* to understanding what happens in many plane crashes. It is why Boeing got into trouble with the 737 MAX.

31

Return for a moment to sticking your hand out of the car window at highway speed. With your hand perfectly parallel to the road, the air hits your thumb but not your palm. It is not easy to keep it still, but it is not very hard either. Now tilt your hand up slightly, giving it a positive angle-of-attack. It becomes much harder to hold it in place, because air hitting your palm produces that Newtonian deflected-air upward force. Your palm is generating *lift*. Give your palm a greater angle-of-attack – tilt it up more – and it becomes harder to keep it motionless because now more air is hitting it and so the upward force – the *lift* force – is even greater. The more you tilt your palm, the more you increase the angle-of-attack and the more *lift* your palm generates.

Same for planes: the more a pilot increases the wing's angle-of-attack, the more *lift* the wings generate. Again, the increased *lift* causes a plane to rise. Pilots increase or decrease their wings' angle-of-attack by raising and lowering their plane's nose. We'll describe how in the next chapter.[12]

Assume air traffic control asks your pilot to slow down but stay at the same altitude. [FIGURE 8] You can hear your pilot reducing engine power, but that reduces *lift*, which we just said would now be overpowered by *gravity* and so your plane would descend. But if at the same time your pilot tilted the plane's wings upward, raising their angle-of-attack, that would increase *lift*, compensating for the *lift* lost by the slower airspeed. Now your plane will stay at the same altitude.

Staying at an altitude when slowing down by raising the angle-of-attack works up to a point. If a pilot raises the angle-of-attack too much, the wings will quit generating *lift* entirely. They will *stall*, and the plane will drop like a rock. It is binary. One second the wings are generating *lift*, and the next, if the angle-of-attack is just a fraction too high the wings will quit flying. The speed at which the angle-of-attack becomes too great is the *stall speed*.[13]

A common misperception is that a stall is when the engine stops. Not true. It seems people equate an airplane's stall with their car's stall on

---

12 A plane climbs or descends when its nose is pointed up or down not only because its wings are at a higher or lower angle-of-attack and generating more or less *lift*, but also because of the logical reason: it is pointed that way. Aerodynamics is a complex subject and we are barely scratching its surface.

13 Planes will stall at any speed, but only at one specific angle-of-attack (the stalling angle-of-attack changes when flaps and slats change the wing's camber line – that explanation is a few paragraphs up). Most of the time that angle-of-attack is reached in normal flying when trying to hold altitude as airspeed drops – like in my example. Since pilots generally watch airspeed closely – or at least they're supposed to – they are taught to focus on airspeed to avoid stalling.

the shoulder of a highway. They have nothing to do with each other. An airplane's *stall* is not about the engine. It is about when the wings are no longer generating lift.

Another misperception is that a stall is a death sentence. Also not true. Pilots usually have a way out of it – a way to *recover* from it.

The first step in recovering from a stall is recognizing it is about to happen. A stall should never surprise a pilot because as the plane nears *stall speed*, it buffets (an aviation term meaning *vibrate*), a horn might go off in the cockpit, and sometimes the yoke actually shakes. Called a *stick shaker*, it makes a loud rattling noise while shaking. Not all yokes shake, and not all planes have a stall warning horn. But every plane buffets, since every stall is caused by similar aerodynamic forces acting on the wings. The other warnings are triggered by sensors when the stall is very close by.

Recognizing an approaching stall – the buffets, horns and stick shakers – is pounded into student pilots. But if their plane steps over that line and stalls, they are taught and drilled on *recovery procedures* – how to pull out of it.

To recover from a stall, the two immediate needs are to reduce the angle-of-attack and to gain speed. Pointing the nose down – though not straight down – does both. It obviously lowers the angle-of-attack because the nose had been pointed too high when the plane stalled. And the plane will gain speed because, like a sled going downhill, all planes accelerate when pointed downward. Raising thrust by pushing the throttles to the firewall also adds speed. Those two pilot moves – nose down and power up – must be immediate and instinctive, so they're practiced again and again.[14]

The angle-of-attack at which a plane will stall – and its stall speed – depends heavily on the shape of its wings. The more pronounced the camber, the more *lift* a wing can generate, and therefore the higher the angle-of-attack – and slower the speed – it can fly with before stalling. Since nothing good comes for free (usually), the cost of *more camber/more lift* is that the wing will also produce more *drag* – more friction with the air – meaning it will need more engine power to overcome that air friction and fly.

A plane flies slowly on takeoff and landing, and fast while at high-altitude cruise. It can do both by changing the shape of its wings during flight, adding

---

14 While recovering this way works for most small planes, it does not necessarily work for jet fighters and some larger passenger airliners. Every plane requires the angle-of-attack be lessened in order to 'break' the stall, and getting the nose down does that. But not every plane needs power added. Some even recommend reducing power initially. Whatever the proper recovery procedure, that is what pilots practice.

camber when it is flying slowly, and removing camber when it is flying faster. *Flaps* and *slats* do that. [FIGURE 9]

Flaps extend out and down from the back of a wing, while slats slide out and down from the front. A single lever in the cockpit operates both in unison. Flaps emerge in stages, measured in degrees – 10°, 15°, 20° and so on are typical. Each higher number means more camber, a longer camber line and the ability to fly more slowly, but also more drag.

Large commercial jets use flaps and slats for takeoff. Once airborne and going fast enough, the pilots retract them until they are needed again for the slowdown before landing. If you've looked out of your plane's window at the wing during takeoff or landing, you've almost certainly seen them.

Chapter 10

Control Surfaces

Now that we know how a plane stays in the air, the next topic is how it moves about. *Control surfaces* make a plane turn, climb and descend. They change the shape of the wing or the tail surfaces, making them generate more or less *lift*. They also deflect air up or down and left or right. [FIGURES 9 and 10]

A plane banks because of *ailerons*, control surfaces on the far end of each wing. [FIGURE 10] Once banked, a plane will turn. To bank *left*, tilting the right wing's aileron down makes the right wing's camber greater than it was, plus air is now deflected downward, which together push the right wing upward.

Simultaneously the reverse happens on the left wing. Its aileron tilts up, reducing camber. Air deflected upward from the up-tilted aileron adds its Newtonian-opposite-force on the wing, and together they push the left wing downward. The right wing has gone up, the left wing has gone down, so now the plane is banked left and it will turn left.

Roughly the same explanation applies to how planes climb and descend, and how pilots increase and decrease their plane's angle-of-attack. Since all the accidents we will be talking about involve climbs and descents, I'll go slowly.

The *horizontal stabilizer*, the small wing in the rear – the tail – has a control surface called the *elevator* that tilts up and down. [FIGURES 9 and 11] With the elevator in neutral – perfectly centered – the horizontal-stabilizer-elevator combination has no camber and isn't deflecting air in any direction.[15]

---

15 The combination horizontal-stabilizer-elevator is rarely perfectly centered, but in explaining their function it helps to say so. For a plane flying straight-and-level it is close to centered, but its actual position depends, among other things, on airspeed and the distribution of weight around the plane. Since I am issuing disclaimers, here is one more. Some horizontal stabilizers have slight *negative* camber while in neutral (that is, pushing the tail down and the nose up), but for our purposes, we can stay with the assumption the horizontal stabilizer has no camber.

For a plane to descend, the elevator tilts down. That gives the combination horizontal-stabilizer-elevator camber, and also deflects air downward. Together they increase *lift* in this tail control surface. The increased *lift* makes the plane's tail go up, which makes the nose go down, and the plane will descend.

For a plane to climb, the elevator tilts up. Camber is now on the underside (the opposite of a wing's camber), and air deflection is upward, so *lift* pushes downward. The tail goes down and the nose goes up, and the plane will climb (the force is still called *lift* even though it is pushing down).

One last control surface. The vertical part of a plane's tail is the *vertical stabilizer*. The *rudder* is attached to it. [FIGURE 11] It isn't obvious when watching an airplane turning in flight, but its tail can be too high or too low relative to the rest of the plane. If the tail is too high the plane is said to be *skidding*, like a car's rear wheels skidding to the outside on a sharp corner. If the tail is too low the plane is described as *slipping* inside the turn – similar to 'understeer' in a car. The pilot uses the rudder to prevent skidding or slipping.

A pilot uses foot pedals to move the rudder. As you might expect, the left foot pedal pushes the rudder's control surface left, which shoves the tail right, directing the plane's nose to yaw left (*yaw* is when the nose veers left or right). The right pedal does the same in that direction.

To move the ailerons and elevator, pilots flying most commercial and general aviation planes have a *yoke*, shaped like a half-wheel and attached either to the instrument panel or to a moveable column or pedestal. [FIGURE 18] Fighter jets and a few small planes (like aerobatics planes) use a *joystick*, a stick between the pilot's legs attached to a universal mount on the cockpit floor, giving it 360° freedom of motion. A few planes, such as Airbus airliners, have a small joystick off to the pilot's right side (or left side for an airliner's captain), called, appropriately enough, a *sidestick*. [FIGURE 17]

They all do the same thing. Pushing them forward tilts the elevator down, causing the plane to descend. Pulling them back tilts the elevator up, causing the plane to climb. Pushing the joystick/sidestick left or right, or turning the yoke left or right (like a steering wheel), tilts the ailerons and banks the plane in the desired direction.

I am going to introduce one more control surface topic now which will come up in the 737 MAX crashes, and then talk about it in more detail when it is most relevant.

Most control surfaces can be *trimmed*. It has nothing to do with taking a little off the sides and top. It is about making it easier for pilots to control their aircraft. *To trim* means to set a control surface so it remains in whatever position the pilot put it in.

An example will make that last sentence clearer. You're piloting a plane at 2,000 feet over your home town on a summer afternoon and want to climb 5,000 feet higher, up to 7,000 feet, to get a better view. Having read this far, you know what to do: you pull back on the yoke, raising the elevator, which points the nose up (you would also add power, but I am ignoring power here). Now you're climbing. If you let go of the yoke the elevator will return to neutral and the nose will drop back to straight-and-level, so you will have to keep pulling the yoke back for ten minutes if you are climbing at 500 feet per minute (500 feet per minute *multiplied by* 10 minutes = 5,000 feet of climb). That is a long time. It is tiring. Instead, you would *trim the plane* so the nose stays raised. Pilots say you would 'trim nose-up.' Now you can relax, loosen your grip on the yoke or even take your hand away, and the plane will climb on its own.

Elevators are trimmed one of two ways: either by a *thumb switch* – a rocker switch on the yoke controlled by a pilot's thumb – or by a *trim wheel* located by the pilot's thigh, next to the throttles. A pilot would push the thumb switch up to trim nose-up, or down to trim nose-down. Alternatively the pilot would turn the trim wheel one way or the other to get the same result.

In jet airliners, either action tilts the front edge of the horizontal stabilizer in the tail up or down [FIGURE 12].[16] The stabilizer normally meets the air cleanly, without an angle-of-attack, but when the pilot uses the thumb-switch or trim wheel to tilt up the front of the horizontal stabilizer, airflow hitting its underside pushes the tail up, and therefore the nose down. The opposite happens if the horizontal stabilizer is tilted down. When the pilot lets go of the switch or wheel, the stabilizer stays in place and the plane will climb or descend precisely as it had when the pilot was pulling or pushing the yoke.

That's it for the basics. It is enough to understand the mistakes some pilots have made.

---

16 On most single-engine propeller planes, a small moveable surface on the elevator, called a *trim tab,* works with the thumb switch or trim wheel to have the same effect as moving the horizontal stabilizer of a large passenger jet. Elevator trim tabs are generally not used in larger aircraft.

Chapter 11

Autos

Aviation automation took off in the decades following the Second World War (excuse the pun), as advances in electronics and computing led to advances in cockpit avionics. *Avionics* is the all-encompassing term for aviation electronic components; I'll use it from now on.

Technology reached a point where if a pilot could see it, so could a sensor, if a pilot could move it, so could a servo, and if a pilot could calculate it, so could an on-board computer. Eventually nearly every function a pilot could execute in the air could be automated. And so, over time, they were.

Not only were autopilots becoming more sophisticated and capable, two new *autos* appeared in the sky: *autothrottle* and *autoland*.

Beginning in the mid-1950s, automatic throttles – autothrottles – put engines under electronic control. The pilot instructs autothrottles to do one of two things: either maintain a specific amount of *power* (or *thrust)*, or hold a specific *airspeed*. Autothrottle can be set to, for instance, takeoff power or climb power, and autothrottle will hold that power level. Or the pilot can instruct autothrottle to hold a specific airspeed, and no matter what the plane is doing, autothrottle will ensure the engines generate the right amount of power to stay locked on that speed.

Like autopilot, autothrottle's great value is as a workload-reducer. A pilot can focus on flying the plane instead of both flying the plane *and* handling engine settings or worrying about airspeed. That is the theory at least. But as we will see, it also adds layers of complexity that has gotten pilots into *a lot* of trouble.

Another workload-reducer, but also another level of cockpit automation, is the Flight Management System, the FMS. It does what its name implies: it is an all-encompassing system for managing a flight and handling the most important housekeeping chores. Pilots use its keypad to enter into their FMS the routes, altitudes, radio beam names and waypoints for every leg of their trip.[17] Once properly programmed, for each step of the journey the FMS sets navigation radio frequencies and works with the autopilot to turn the plane to new headings and climb and descend to new altitudes, faithfully following its programmed route until the pilot takes over, usually during the approach to an airport.

More than once I have sat on a plane on a taxiway while the pilot entered a new routing into the plane's FMS to avoid bad weather that suddenly appeared along our flight path. It takes a while, but with automation doing so much of the flying, there is no other way. You would prefer your pilots taking their time loading the FMS. Mistakes have happened – wrong waypoints have been entered – sometimes with fatal results.

The FMS and autopilot do not have to relinquish control of the plane to the pilots during the last phase of a flight, when they are approaching the runway. If properly programmed and if the plane and runway are both designed for it, automation can land the plane using *autoland*.

If autopilot and autothrottle seem a bit like magic, *autoland* is positively spooky. *Autoland* will guide a plane all the way to touchdown, enabling landings under what would otherwise be impossible visibility conditions. It is really part of the autopilot, though a part that is only useable when everything is lined up for it.

Around half of all fatal commercial plane crashes happen between the descent and the landing, making it by far the most dangerous phase of a flight (three times as much as takeoff and the initial climb-out, the second-most dangerous phase). Not all landing-phase crashes occur in bad weather, or even darkness. Some have occurred on gorgeous blue-sky days when pilots failed to execute what they had been trained to do since the very beginning of their flying careers. Anything that makes landing safer is welcome. Autoland is one of those things.

In the next chapters we are going to discuss three crashes that happened during landing. To truly understand what went wrong, and automation's role in the accidents, we have to spend a few minutes reviewing how an airplane returns safely to earth.

---

17 Waypoints are designated destinations on a route. They might be the intersection of two navigation radio beams, or the antenna location of a single beam, or even an imaginary point on the ground that only GPS can identify. They are marked on aviation navigation maps. Pilots plan routes from waypoint to waypoint, from their originating airport to their destination.

Chapter 12

Landing

All planes land the same way. The pilots of an Airbus A380 with 550 people aboard do roughly the same things in the cockpit to land that behemoth as they did while learning to fly in a 2-seat training plane. Pilots on big jets have more things to do and less time to do them, but the basics are identical.

The last thirty miles or so of a plane's path to its destination runway is called the *approach*. The last five to ten miles, when the pilots have the runway lined up in front of them, is the *final approach*.

For commercial planes landing at major airports, there are two varieties of final approach: *visual* and *instrument*.[18]

During nice weather and good visibility, airline pilots often fly a *visual approach*. As its name implies, since they can see where they are going, pilots are expected to use their eyes and three-dimensional judgment to guide their plane to the runway. Pilots can hand-fly a *visual approach*, or they can use the autopilot to fly it for them until they are a few hundred feet above the runway. But honestly, since most pilots want to *fly* their planes, they usually jump at the chance to hand-fly 'the visual.' While *on the visual* they can use cockpit instruments to ensure they are on target, but, as we will see, that does not always keep them out of trouble.

When visibility is poor – sometimes at night, always when clouds are lower than 1,000 feet above the ground – pilots fly an *instrument approach*, where they use their instruments to follow two narrow radio beams down to the runway threshold. The two beams are the *localizer*, and the *glide slope*, and a single cockpit instrument tracks them both. [FIGURE 13] That instrument is what pilots *on the visual* might use to ensure their approach is lined up perfectly.

---

18 In commercial aviation, both of these approaches are flown under what the FAA calls *Instrument Flight Rules – IFR* – the rules for flying in clouds, as well as when flying above 18,000 feet. The other set of rules are *Visual Flight Rules – VFR* – which apply when pilots can see a minimum distance and are well away from clouds.

The *localizer* beam is aligned with the center of the runway. On final approach the cockpit instrument tracking the localizer tells the pilot if the plane is to the left of the runway (and if it is, how far left), to the right of it (and how far right), or heading straight down the middle. It is like someone guiding you into a parking space at the mall so your car ends up right between the lines.

At the same time, the *glide slope* beam lays out, on a steady downward slope, a perfect descent path to the runway. [FIGURE 14] That perfect path is roughly a 3° angle up from the runway (or down to it). While flying that descent, the same cockpit instrument tracking the localizer beam lets the pilot know if the plane is above the glide slope beam and how far above, below it and how far below, or smack on it.

Instrument approaches using both of these beams guiding the pilot are called *Instrument Landing System approaches*, *ILS approaches* for short. Some approaches only have the localizer beam, and are called, logically enough, *localizer approaches*. No approaches only have a glide slope.

Pilots on ILS approaches can't follow the glide slope beam all the way to the ground because its accuracy degrades below 200 feet of altitude (at this height a commercial jet would be approximately three-quarters of a mile from the touchdown point). Even in this modern electronic era the glide slope signal isn't perfectly reliable below that height because of potential interference from nearby buildings, other planes, and even vehicles on the ground.

At 200 feet the pilot must make a decision: land, or *go around*. Since a pilot can't land on what he can't see, if the runway isn't visible the pilot must immediately abandon the approach: put in full power, climb and *go around* the airfield to try again, or fly to another airport.[19]

Airports in England are slammed by fog and rainy weather for much of the year. An altitude of 200 feet is frequently not low enough for pilots there to see the ground and land their plane. Because of this, in the 1940s and '50s English engineers took the lead in inventing *automatic landing* (like the other *autos*, autoland started as two words).

Autoland marries the autopilot with the avionics tracking the two radio beams on an ILS approach so a plane's computers follow the beams

---

19 A couple of comments. First, the height when a pilot must see the runway or go around, called the *decision height*, isn't always 200 feet. Sometimes it is higher (but it is almost never lower). Second, ILS radio beams are still used by most of the aviation world, but GPS-based systems are also in use, and are rapidly supplanting ILS to be the dominant form of instrument approach. They don't use radio beams, but software that maps out the course to landing *as if* the beams were there. One advantage of GPS is that radio beams are straight lines, but GPS-based approaches can have turns in them.

all the way to the 200-foot decision altitude. To that point it is no more sophisticated than a good autopilot, but getting down the last 200 feet makes autoland special.

Since the glide slope beam is not trustworthy below that height, avionics aboard the plane merges whatever it can detect of the radio beams with its *inertial navigation system* and a *radio altimeter* to fly the last 200 feet. Inertial navigation systems track a plane's position without outside help (like GPS). A radio altimeter, also called a *radar altimeter*, bounces a radio signal off the ground under the plane to determine the plane's height within *inches*. That accuracy is necessary for autoland to know the exact instant to execute the plane's *landing flare*.[20]

No matter how a plane approaches the runway – whether on a visual or instrument approach, being hand-flown or on autoland – the *landing flare* is how all planes descend the last few feet. That is the last thirty or forty feet for an airliner, the last ten for a small plane. It is both a verb and a noun. To *flare* the plane is to tilt the nose up and reduce power to *idle* (that is, to nearly off, just like when your car engine *idles*). Then the plane stays in *the flare* – in that nose-high attitude with the power at idle – until the main wheels touch the ground.

A plane flares to slow its rate-of-descent, and to position the main landing gear to hit the runway first. The nosewheel would collapse from the shock of the plane's weight returning to earth if it made contact first.

While *in the flare*, pilots sense what feels like a cushion of air forming under their wings, called *ground effect*. It is not really an air cushion. They are actually feeling lessened aerodynamic drag and improved *lift*, and their plane seemingly floats as it slows until settling on the runway (though we all know some touchdowns are more floating than others). The landing flare has to occur at a very specific point above the ground. If executed too high, a plane could stall and nosedive into the runway. Too low, and a plane will miss ground effect and hit the runway too hard.[21]

---

20 To expand a bit, an *Inertial Navigation System* uses on-board gyroscopes and accelerometers, combined with airspeed and altitude data, and with no outside help, to figure out precisely where the plane is at all times. It is first given a starting point ('You Are Here'), and from there tracks the plane's every movement, updating its position continuously. It is very accurate over thousands of miles, so it can handle the plane for a few seconds at the end of an ILS approach. GPS alone is not accurate enough.

21 A raised nose doesn't guarantee the cushioning sensation of ground effect. Navy pilots landing on aircraft carriers have their noses up but fly through the ground effect – they *fly the plane* to the runway – by keeping their rate-of-descent high and smacking onto the carrier deck on landing gear designed to take the punishment.

An obvious question is, if planes can autoland, can they auto-takeoff?

Logically they can. Takeoff is simpler than landing, *simpler* being a relative term. On takeoff the pilot pushes the throttle forward to put in power, uses the rudder pedals to keep the plane in the middle of the runway as it accelerates, and at the right speed pulls back on the yoke, joystick or sidestick. Presto, the plane is now flying.

Landing, on the other hand, requires skill and finesse. It is really an athletic event, with the pilot using eye-hand coordination and judgement borne of experience in equal measure to descend from altitude to the perfect point to flare, and then making infinitesimally small adjustments in those last seconds while sinking slowly in ground effect to produce a greaser of a landing, one where you would not know you've arrived unless you looked out your window.

That is why I made the point in the beginning of our story about proficiency and practice. Pilots out of practice won't land a plane as well as those who have done it repeatedly and recently.

Auto-takeoff is not used for a couple of reasons. First, it is illegal. In the US, the Federal Aviation Administration – the FAA – which sets the rules for planes and pilots, does not allow crews to engage autopilots below 500 feet. Most large airlines are granted exceptions allowing their pilots to engage autopilot at lower altitudes, but no airline's pilots are allowed to begin the takeoff roll with their hands in their laps. The FAA wants a pilot's hands on the throttle and yoke or sidestick, ready to react instantly in case something goes wrong.

Which leads directly to the second reason. Pilots believe – rightly – they and their plane are safest in the air, the higher up the better. If trouble strikes, altitude gives pilots room to figure out a solution. But on takeoff they have no altitude to work with. If something goes wrong – engine failure, blown tire, bird strike – a pilot needs to react instantly and perfectly: solve it quickly, slam on the brakes equally quickly, or get the plane into the sky where the problem can be dealt with. As with stalls, it is something pilots train for incessantly. They don't – well, most of them don't – trust their autopilot to decide for them what just happened and what to do next. Pilots usually want their lives in their own hands when risks are high.

Knowing what I know today about aviation automation, I don't blame them.

Part II

Hands On

Chapter 13

Sniffles

As planes have flown faster and with more passengers, the consequences of human error have multiplied. Automation has reduced the chance of errors occurring, but it has not eliminated them. My education is in computer science. The first thing we learned as college freshmen was a computer is only as good as its software and the information entered into it. It will not do what it hasn't been programmed to do. It will not react to information it does not have. It *will* react to information it has been given, even if it was given in error.

Once a pilot programs a route into the Flight Management System, the plane will follow it precisely. Mis-type the code letters for a waypoint along the route, and the autopilot will take its plane miles off course. The FMS can't know what the pilot was thinking, and so it can't find or fix a mistake. A human has to spot the error and make the change.

The same goes for switches and buttons. Even the most automated airliner still requires its pilots to push, pull and turn buttons, dials, and toggles. Every setting has meaning.

But what if the pilot does not realize the consequences of one of those pushes or pulls or turns – or just forgot? The switch will do what it is supposed to do, regardless of whether the pilot remembered. Then what if, when the switch does its thing, the pilot is so shocked, so surprised, that he causes an otherwise perfectly good airplane to crash?

That happened aboard a flight on a snowy winter evening in February 2009, the continuation of Colgan flight 3407 from Newark to Buffalo's Niagara International Airport, whose takeoff I described in Chapter 1. The flipping of a single switch led to the deaths of forty-nine people on that plane and one on the ground. It is not the only reason the plane crashed, but it may be the most important.

Shenandoah, Iowa, is a town of around 5,000 people around 100 miles southwest of Des Moines. Marvin Renslow was born there in 1961.

He played the drums in high school – as an adult he still occasionally banged away on a set in his home – and earned a degree in aviation from Guilford Technical Community College in 1992. Along the way he got his private pilot license and instrument rating. Twelve years later he began serious flight training with the goal of launching an airline career, attaining that goal with his hiring by Colgan Air. Eventually promoted to captain, when his schedule permitted he was a regular at his Baptist church near Tampa, Florida, along with his wife and two teenage boys.

Colgan Air is one of those companies you might not have realized exists in aviation. When you board a Delta Air Lines plane at New York's LaGuardia Airport bound for Chicago, sometimes you are actually getting on a Republic Airways plane painted in Delta's livery and assigned a Delta flight number. Other Republic Airways planes are flying under United Airlines and American Airlines colors. It is one of the ways the majors manage their overhead – by contracting out their shorter routes to companies like these.

That was Colgan Air's business until closing its doors in 2012. Colgan flew Bombardier Q400 and Saab 340 twin-engine turboprop commuter airplanes all over the east coast on short-haul passenger routes for US Airways, United and Continental.

On the Thursday night of 12 February 2009, Captain Renslow commanded Colgan flight 3407, a Bombardier DHC-8-400 sometimes called the 'Dash-8' because it was originally designed by de Havilland Canada, but more commonly known as the Q400 ('Q' for the Q variant). Flight 3407 was a Continental Connection trip. Renslow occupied the cockpit's left seat, the captain's position. In the right seat next to him was his copilot, Rebecca Shaw.

Shaw grew up in Maple Valley, Washington, a small town south-east of Seattle, where she played volleyball. From her senior year in high school she set her sights on an airline career. To get there, she worked as a flight instructor while attending Central Washington University, graduating in 2007 with a degree in flight technology. She spent the next year as a flight instructor in Phoenix, Arizona, until Colgan hired her. She was married and had recently moved back to Maple Valley.

Renslow had slept in Newark the night before the crash, though no one is sure exactly where. He didn't have a crash-pad – as they are perhaps inappropriately called – near the airport. Nor could investigators find a record of him sleeping in any nearby hotel. That morning an early-arriving Colgan captain spotted him asleep in the airline's crew room, which was not permitted but done anyway by crewmembers with flights out of Newark later in the day. The point is he may not have gotten a good night's rest.

The same questionable night's sleep might have also been true for Shaw. It is not uncommon for flight crewmembers to live far from the airport they normally fly from. Shaw lived clear across the country, so she had hitched a pair of overnight rides the evening of 11 February, going first to Memphis, Tennessee, and then just before dawn on to Newark. Reportedly she slept for at least part of the journey. And during the day she texted someone that she had managed to take a nap. She was also fighting a cold.

It is likely neither pilot was at their best. But both were experienced professionals with thousands of hours in their logbooks, and on the day of the flight nothing obvious indicated impending trouble.

After any commercial airplane crash, the NTSB, the National Transportation Safety Board, the US government agency charged with investigating all significant transportation vehicle accidents, routinely analyzes the plane's Cockpit Voice Recorder. The CVR, required on all commercial aircraft, contains every sound heard and word spoken in the cockpit during the two hours before a crash. A transcript is usually included in the NTSB's final accident report. Colgan 3407's CVR transcript is particularly illuminating.

The flight was supposed to take off at 7:10 that evening, but didn't even push back from the gate until 7:45, and it was another 45 minutes until they began taxiing to the runway (all times for Colgan flight 3407 are Eastern Time). The delay was caused by high winds in the area, which was backing up the schedules of all the airlines flying into and out of Newark. While waiting in their seats, the plane's CVR recorded Renslow and Shaw discussing a host of things, including flight crew schedules, getting airborne in hot weather (hot weather concerns pilots because it requires a longer takeoff run than cooler temperatures), their salaries, and the Seattle housing market.

Perhaps most germane, during their conversation Shaw complained to Renslow of not feeling well. Since the tape began she had sneezed at least once and sniffled a few times. At 8:41, less than thirty minutes before takeoff, they had this exchange:

Shaw: *Oh I'm ready to be in the hotel room.*
Renslow: *I feel feel feel bad for you as far as feeling [unintelligible].*
Shaw: *Well this is one of those times that if I felt like this when I was home there's no way I would have come all the way out here. But now that I'm out here.*

Renslow: *You might as well.*
Shaw: *I mean if I call in sick now I've got to put myself in a hotel until I feel better. You know we'll see how how it feels flying. If the pressure's just too much I you know I could always call in tomorrow at least I'm in a hotel on the company's buck but we'll see. I'm pretty tough.*[22]

Then one of them sniffled. The CVR could not discern who.

---

22 All cockpit conversations in the book are from CVR transcripts in government accident reports. I edited punctuation only if it was necessary to clarify meaning.

Chapter 14

One Switch

In commercial airplane cockpits, traditionally one of the two pilots has their hands on the controls doing the flying. In accident reports and among air crews, that pilot is called the Pilot Flying. The other pilot handles radios, reads checklists aloud, operates the flaps, slats and landing gear, communicates with passengers and flight attendants, monitors airspeed and altitude, keeps an eye on the engines, and generally takes in everything going on. That pilot goes by one of two names, both accurate descriptions: either the Pilot Not Flying, or the Pilot Monitoring. I will use the Pilot Not Flying.

Cockpit crews on long flights might switch responsibilities half way into the trip – one will fly the takeoff, the other the landing. On short-haul routes where a cockpit pairing may spend an entire day together, they usually switch off flying and monitoring legs. On this flight to Buffalo, Marvin Renslow would be the Pilot Flying, the one controlling the plane from takeoff to landing, with Rebecca Shaw, the Pilot Not Flying, handling the other duties. Conforming to those duties, while in the line of planes waiting to take off she diligently read, and with Renslow's participation completed, the appropriate checklists.

Finally, at 9:18 pm, Colgan 3407 got into the air. A few minutes after takeoff, while climbing up to ten thousand feet, Renslow turned on the autopilot. At some point around then he also turned on de-icing equipment including – most crucially – setting a switch labeled **REF SPEEDS** to INCR, increase. The CVR did not record Renslow commenting aloud as he flipped the de-icing switches. But the NTSB Accident Report says the Flight Data Recorder – which records the positions of control surfaces and switches throughout the flight and is analyzed during a crash investigation – revealed the switches were set while the plane climbed away from Newark. The de-icing switches were on a cockpit overhead panel directly above Renslow, so he probably set them.

THE DANGERS OF AUTOMATION IN AIRLINERS

REF SPEEDS is short for *reference speeds*. When set to INCR, the plane's stall warning goes off at a lower angle-of-attack, meaning a higher airspeed. The warning speeds are increased by 15 to 25 knots, depending on the degree of flaps deployed for landing. It does *not* mean the plane will stall at a higher speed. It only means the *stall warning* will go off in the cockpit at a higher airspeed. For planes flying into icing conditions, it could be a lifesaver.

Icing is the accumulation of ice on a plane's wings, tail and fuselage, adding weight and changing the camber of the wings. In small quantities it has little or no effect, but if its accumulation is thick enough, icing makes controlling the plane more difficult and can be very dangerous.

While sitting on the ground, ice can build up during freezing rain or snow. It is removed before takeoff by spraying it with a heated liquid-chemical mixture from a high-pressure hose. In the air, planes can encounter icing while flying through any moisture near freezing temperatures. That usually means in clouds, but it could also come from rain or snow below clouds.

Pilots prefer to not fly in areas where meteorologists report the potential for icing. But if it cannot be helped, commercial airplanes have ways of getting rid of it, or to at least try. Colgan's Q400s had electrically heated propellers and windshields, and 'icing boots' that physically knock ice off the wing-edges and tail surface edges. These systems are not guaranteed to work, so avoidance is best.

All pilots justifiably worry about icing. Some, especially those who have never before encountered it in flight, positively fear it. A pilot who knows ice is accumulating on his airplane – or *thinks* it might be – will fly faster than normal on approach and landing because he would expect his plane to stall at a higher airspeed (and lower angle-of-attack) than when free of ice. That is why the designers of the Q400 installed the **REF SPEEDS** switch, in effect raising the stall *warning* speeds when its pilots turned on the de-icing switches.

Colgan 3407 was scheduled for 53 minutes in the air, plenty of time for Renslow and Shaw to continue the wide-ranging conversation they had begun on the ground. Around 9:51, while they were chatting and cruising at 16,000 feet, Shaw used ACARS, an air-to-ground communication link with their airline's operations center, to find out their landing speed. As the Pilot Not Flying, that was among her responsibilities.

ACARS, the Aircraft Communications Addressing and Reporting System, is how airline operations centers and flight crews stay in touch with each other. Its two-way communications cover everything from the mundane to the vital, including en-route weather, push-back clearance, takeoff times, and landing speeds.

Aboard Colgan 3407 Shaw radioed her operations center with pertinent information about their flight, and in two minutes she got a response, relaying to Renslow that their landing speed would be 118 knots. Renslow acknowledged he heard and understood.

Every pilot knows at least roughly how fast to fly at touchdown. On small planes used for flight training, student pilots memorize that speed, which varies with the amount of flaps their plane is using. In large passenger planes that speed also varies with the amount of flaps and slats, as well as with air temperature, the airport's altitude, and the plane's weight, to name a few extra factors. Those factors affect small training planes as well, but they usually don't have enough impact to change memorized landing speeds.

The landing speed ACARS gave Shaw, 118 knots, is approximately the normal landing speed for a Q400 with passengers and luggage aboard and using 15° of flaps. Since airplanes land a few knots faster than their stall speed, the 118-knot figure meant the plane's *stall speed* would be around 110 knots.

Renslow clearly knew the 118 knot landing speed was in the right ballpark because he acknowledged the number without comment. Same with Shaw.

But both of them should have realized something was wrong.

Knowing his airplane as any pilot should, Renslow must have understood that turning the **REF SPEEDS** switch to INCR changed the plane's stall warning to now trigger at around 130 knots – well before they'd reach 118 knots. When Shaw mentioned the 118-knot landing speed, Renslow should have remarked on this and made a decision: either they were going to land at 138 knots – 20 knots faster than the speed Shaw gave him – or he needed to turn off the **REF SPEEDS** switch.

He did neither.

Shaw should have realized the same thing. She should have remembered Renslow had turned on their de-icing equipment and flipped **REF SPEEDS** to INCR and so she too should have known the stall-speed warning would go off before they reached 118 knots. Even if she hadn't seen Renslow set **REF SPEEDS**, she knew they were flying in icing conditions. That alone should have clued her into the need for higher landing speeds. Either the speed from ACARS needed to be changed, or the switch should have been turned off.

She proposed neither.

Renslow's and Shaw's inaction here was a mistake.

Three minutes later, at 9:56, Shaw asked Renslow to begin their descent into Buffalo earlier than normal so the air pressure changes in the plane would be gentler on her cold-stuffed head. As she explained, 'Might be easier on my ears if we start going down sooner.'

'Yeah,' Renslow agreed, 'we could do it. That's fine.'

While flying within the Air Traffic Control system, ATC, planes are not allowed to change altitude, heading, and sometimes speed, without permission. So Shaw made the request of ATC, which allowed them to descend to 11,000 feet.

Since the autopilot was controlling the plane, Renslow, the Pilot Flying, was not hand-flying. Instead he was 'flying' by instructing the autopilot to take them higher or lower, faster or slower, left or right, by twisting knobs and pushing buttons on the instrument panel to give the autopilot new airspeed, altitude or heading targets. He now dialed **11000** into the autopilot, and it began taking them down.

Meanwhile Renslow and Shaw continued chatting away. At 10:05 ATC cleared them to descend to 6,000 feet, meaning the FAA's *Sterile Cockpit Rule* was about to go into effect. The rule orders pilots not to speak about anything not directly pertaining to their flight when below 10,000 feet during crucial phases of a trip. That includes takeoffs and landings.

With the Sterile Cockpit Rule now governing, Shaw and Renslow got serious, talking only about details of the approach they were about to fly – the ILS Approach to Buffalo's Runway 23.

Their cockpit would not remain sterile for long.

Chapter 15

Icing

At 10:10 that evening, five minutes after the Sterile Cockpit Rule went into effect and now at an altitude of around 5,000 feet, Shaw launched into an appropriate conversation about icing:

Shaw: *Is that ice on our windshield?*
Renslow: *Got it on my side. You don't have yours? [Renslow whistles]*
Shaw: *Oh yeah oh it's lots of ice*
Renslow: *Oh yeah that the most I've seen – most ice I've seen on the leading edges in a long time. In a while anyway I should say.*
Shaw: *Oh [unintelligible]*

While that dialogue was acceptable, Shaw kept going, telling Renslow all of her previous flying experience before Colgan was around Phoenix, Arizona, where she had never confronted icing. While that is mildly interesting, it had nothing to do with their present task, namely landing safely in Buffalo. She shouldn't have taken their conversation there.

Still, Renslow took her cue. As the plane's autopilot continued flying the approach and Shaw handled air traffic control radio calls, they stayed on the topic. Here's some of it, picking up from 10:12 that night. Notice how air traffic control, altitude, heading, and irrelevant personal history are all jumbled together:

Shaw: *I've never seen icing conditions. I've never deiced. I've never seen any – I've never experienced any of that. I don't want to have to experience that and make those kinds of calls. You know I'd've freaked out. I'd've have like seen this much ice and thought oh my gosh we were going to crash.*
Buffalo
Approach   *Colgan thirty four oh seven descend and maintain two*
Control: *thousand three hundred.*

Shaw:	[into her mic] *Okay down to two thousand three hundred Colgan thirty four zero seven.* [to Renslow] *Um two three alt sel. I've got you in pitch pitch hold. I don't know if that's what you want.*
Renslow:	*Two three alt sel. Yeah that's alright, let's uh, we'll do vertical speed back.*
Shaw:	*But I'm glad to have seen oh, you know now I'm so much more comfortable with it all.*
Renslow:	*Yeah uh I I spent the first three months in uh Charleston West Virginia and uh flew—.*
Buffalo Approach Control:	*Colgan thirty four zero seven turn left heading three three zero.*
Shaw:	*Left heading three three zero Colgan thirty four zero seven.*
Renslow:	*Left three three zerooo. We're in heading mode now. Go to blue needles. But I— first couple of times I saw the amount of ice that that Saab would would pick up and keep on truckin'.*
Shaw:	*Yeah.*

Cutting through the exchange is the pilots' preoccupation with icing. How much ice was building up, and how worried should the crew have been? After the crash, NTSB tests showed ice on the wings and airframe totaled between ten and twenty percent of what would be considered dangerous. Their tests also revealed the ice build-up was not impacting handling characteristics at all. The plane continued flying like it would have on a sunny summer day. Icing was a non-factor.

Nearing Buffalo's Niagara International, the two pilots went through the plane's Descent Checklist, then the Approach Checklist. Everything continued looking good.

Two minutes later, flying at 184 knots, Renslow recognized they were moving too fast for this stage of the approach. 184 knots wasn't particularly fast and wasn't even necessarily unusual. But it was time to get their airspeed down closer to landing speed. He asked Shaw to lower 5° of flaps, which would increase the plane's drag and slow it down a bit.

ATC radioed Colgan 3407, ordering them to stay at 2,300 feet until reaching the ILS radio beams, which were a few miles ahead of them. ATC's instructions concluded with, 'Cleared ILS approach runway two three.'

They now had permission, when they reached the ILS beams, to follow them down to the runway. All that was missing was the definitive 'cleared to land,' which would come up in another few minutes.

A light snow was falling.

At 10:16, flying at 2,300 feet and 180 knots, Renslow asked Shaw to lower the landing gear while he reduced power almost all the way to idle. The landing gear would add significantly more drag to slow the plane quickly, while pulling back on the throttle would help slow them as well.

Shaw pulled the landing gear lever as the autopilot automatically raised the nose to keep the plane at 2,300 feet as it slowed down.[23] Presumably Renslow intended to let the autopilot fly the plane until they were a few hundred feet above the ground, when he would take over.

ATC now told the crew to switch their radio frequency to Buffalo Tower, and wished them a good night. Shaw responded by confirming the new frequency, politely adding, 'you do the same.'

Then, as the Pilot Not Flying, Shaw announced the landing gear was down (a cockpit indicator light would have confirmed that). Airspeed was now around 145 knots.

Renslow next asked her to lower the flaps to 15°. Autopilot raised the nose further.

Three seconds later the plane's speed reached 131 knots.

Suddenly the stick shaker went off, its rattling sound surprising everyone.

Then a horn began blaring loudly.

In the Bombardier Q400, as well as in many large commercial aircraft, the main stall warning is a *stick shaker*. The control column actually vibrates – not so much that it can't be handled, but enough so it is unmistakable – and it makes a loud rattling noise, warning the pilots that their airplane's combination of slow airspeed and high angle-of-attack puts it very close to stalling. It does *not* mean the airplane has actually stalled, only that if the angle-of-attack increases a bit more it *will* stall.

If the stick shaker is triggered, the Q400's autopilot is programmed to disconnect immediately, setting off a horn in the cockpit. That is what happened on Colgan 3407.

Renslow reacted almost instantly – in less than a second – meaning his hands were either in his lap or resting lightly on the yoke. He did two things, one right but not right enough, and the other inexplicably and horribly wrong.

First, with his right hand he pushed the throttles forward, increasing engine power, though not all the way to *full power*. That was directionally correct, though he should have shoved the throttles all the way to the stops as he had been trained to do.

---

23 We covered this maneuver in Chapter 9 – raising the nose to hold altitude while slowing down.

He also pulled back on the yoke, first just with his left hand and then with both.

Huge mistake.

That was decidedly *not* what he was trained to do. As we talked about in Chapter 9, the instantaneous, instinctive response to a stall warning of every pilot on Planet Earth should be to lower the angle-of-attack by pushing the yoke or stick forward. Renslow pulled back. And he kept pulling back, harder and harder as the plane's nose rose and speed dropped, and eventually the Bombardier actually did stall. Twenty-seven seconds after the stick shaker went off, the plane crashed, Renslow hauling back on the yoke almost the entire time.

The NTSB report's unintentionally saddest comment was simple and direct. Referring to the moment the stall warning went off and Renslow pulled back on the yoke, the report said, 'The airplane was not close to stalling at the time.'

Chapter 16

Chain Links

Most pilots and every aviation accident investigator who ever lived will tell you accidents never have a single cause. They are almost always the result of a series of events – often very minor and maybe very unlucky events, but a series nonetheless – strung together as links in a chain, resulting in a tragedy. Follow the links and you'll have a clear picture of why the plane crashed.

But blaming a linked chain of events dilutes the most important factors, spreads blame around perhaps unfairly, and might let some*one* or some*thing* truly culpable off the hook. Even when it is unequivocally clear no single event caused a crash, in the end observers can usually identify one blatant act of commission, or maybe of omission, that tipped the scales against the doomed flight.

Applying that to Colgan Air 3407, the plane crashed on the approach to Buffalo that evening because of Captain Marvin Renslow's single act of commission – reacting incorrectly to the plane's stall warning. It also crashed because a series of minor events; small links strung together in the cockpit stuck Renslow in a corner he could not get out of.

The links in the chain leading to Renslow's horrible airmanship mistake were part of a single theme: the crew was not totally engaged in their flight. They had let the Q400's autopilot do the work of flying, allowing themselves to become passengers, along for the ride, chatting away as they descended. With automation – the autopilot – taking them where they wanted to go, the links formed the chain that killed them.

First, they forgot about one switch setting – the **REF SPEEDS** set to INCR.

Second, they forgot the implications of that setting – that it would *automatically*, by itself, change the stall warning speeds (by changing the angle-of-attack warning).

Third, they were deep in conversation.

Fourth, they were unquestionably tired.

And fifth, Shaw was fighting a cold.

Without those links Renslow would not have had the opportunity to make such a serious piloting error.

We also know Renslow and Shaw were looking at their instruments but not really *seeing* them, because reminders about the **REF SPEEDS** switch setting were visible in two places. They missed them both.

Like all modern commercial aircraft, a Q400's instrument panel looks very different from the impossibly busy dial- and gauge-filled cockpits of the early jet age. Renslow and Shaw were looking at a series of flat-panel computer screens arrayed before them. Each pilot had a main screen for flight instruments called a primary flight display, a PFD. [FIGURE 15] The airspeed indicator part of the PFD is digital, with markings on it showing the speed that would set the stick shaker rattling. The NTSB accident report called those markings the *low-speed cue*.

The low-speed cue airspeed was computed by software aboard the plane and would have been close to what ACARS provided Shaw, except it automatically took into account that **REF SPEEDS** had been set to INCR. That meant the low-speed cue was set to 131 knots.

The NTSB calculated that both pilots could see the low-speed cue's markings for 18 seconds before the stick shaker went off. Had they only *looked* at the airspeed part of their PFDs, had they only been paying attention, they might have wondered why the low-speed cue was appearing at such a high airspeed, considering they thought their stall speed was around 110 knots. It is hard to believe neither of them saw it, because Renslow, at least, had looked at his plane's airspeed at 184 knots when he asked Shaw to lower the flaps and landing gear to help slow down the airplane.

Yet somehow they both missed it.

The crew had a second, even more blatant chance to remind themselves about the **REF SPEEDS** switch. The cockpit's engine instruments, presented on a flat-panel screen between the pilots, displayed the words 'INCR REF SPEED' in white capital letters against a black background. There it was, in plain sight. Those words appeared on the screen specifically – and only – because Renslow had hit that switch. Even a casual glance at that engine panel should have sufficed to spot these words.

Yet neither pilot noticed this either.

The NTSB tried finding reasons why the **REF SPEEDS** reminders were missed, but came up empty. In their crash report investigators wrote, 'the failure of both pilots to detect this situation [the approach of the stall warning] was the result of a significant breakdown in their monitoring responsibilities and workload management.'

Renslow and Shaw were either unusually unobservant, preoccupied by something (their conversation, for instance), or they were complacent, content to let the flight move along because the autopilot had things well in hand. *Complacent* is a characterization I'll dive into a few pages ahead.

The NTSB also searched for why Renslow pulled back on the yoke instead of pushing it forward, but again could find nothing. While coming up the ranks, he had failed four separate FAA check-rides for new ratings or licenses, each time requiring a second flight before passing.[24] He also received three 'unsatisfactory' grades from Colgan check pilots during *proficiency check flights*, regular tests of his skills and capabilities. While his grades and retests don't speak well for Renslow's basic airmanship skills, none of those failures were for mishandling stall recovery procedures. In those situations he did fine. He knew what to do. So this can't be added to the links in the causality chain.

In training, Renslow also demonstrated a preference for using autopilot rather than having his hands on the controls in bad weather. Again, that is not why he pulled the yoke back instead of pushing it forward.

It is clear Renslow first panicked and then completely froze, incapable of doing anything besides pulling back on the yoke as the ground rushed up. Meanwhile Shaw's senses were probably overloaded, with the plane gyrating, a horn blaring, and the stick shaker rattling loudly. In the 27 seconds from stick shaker activation to crash, she raised the flaps and landing gear. She did that not because Renslow asked, and not because it was on any emergency checklist. Rather, she did it probably because that is what she learned to do as a flight instructor in Phoenix a couple of years earlier. Those reactions were automatic for her, as automatic as pushing forward on the yoke should have been for Renslow. Her actions didn't help matters and they may even have hurt, though nothing was hurting them as much as Renslow.

No matter how incompetent Renslow may have been as a pilot, he had survived a long time in that left seat because he knew how to get passengers safely from Airport A to Airport B under most conditions. He had appropriately set the **REF SPEEDS** switch on INCR because the plane was flying into possible icing. Then he and Shaw forgot about it.

---

24 Check rides are in-flight tests with a senior pilot, or an airline or FAA examiner, either in a simulator or aboard a real airplane in the air, where certain piloting skills are to be displayed. Pilots take check-rides for new licenses and ratings, and to earn or maintain a particular status (for instance, the right to fly a certain airplane type).

THE DANGERS OF AUTOMATION IN AIRLINERS

In an airplane, even seemingly little things can matter very much.

To me, there is a way the flight might have concluded uneventfully. Had Renslow turned off the autopilot at 10,000 feet, as many captains do, he would have been hand-flying the approach and been more connected with the flight, more aware of what was happening aboard Colgan 3407. He then might have remembered he had set **REF SPEEDS**, or noticed the low-speed cue on the airspeed indicator, or even read the INCR REF SPEED warning on the engine instrument flat-panel display. Anything keeping his and Shaw's tired and head-cold-stuffed minds engaged with their flight would have helped. It is even possible that if Renslow was hand-flying, he might not have been quite so startled when the yoke began shaking, and perhaps would have reacted appropriately, as he always had during his check rides.

Automation had taken them *out of the loop*, as human factors experts would say.

I have proof.

Interestingly, and maybe incredibly, one month after Colgan 3407 crashed, the pilot and copilot of another Colgan commercial flight repeated the identical mistake made by Renslow and Shaw. This plane was landing at night at Burlington International Airport just east of Burlington, Vermont.

Flying the same aircraft type as Colgan 3407, the cockpit crew forgot they had set the **REF SPEEDS** switch in their plane to INCR. Far along on the approach to Burlington, the plane's stick shaker went off unexpectedly, at around 1,800 feet and 135 knots.

This time the plane was not flying on autopilot. Instead the captain – as with Colgan 3407, she was the Pilot Flying – was hand-flying, and she reacted immediately and appropriately to the surprise by adding power and lowering the nose. While doing that, both she and the copilot simultaneously realized their error – they had forgotten about the **REF SPEEDS** switch. Though the captain could have reached up to the overhead panel and simply turned it off, she chose instead to fly the remainder of the approach at higher speeds, as if icing was an issue, and continued with the landing, touching down safely.

The passengers probably didn't notice a thing.

Chapter 17

The Situation

The stick shaker incident aboard the Burlington-bound Colgan flight that did not crash shows why automation is not only an important part of the cause, it is perhaps the *entire* cause of the loss of Colgan 3407. With her hands on the yoke and throttle as they approached Burlington International, the captain of the Vermont Q400 was fully engaged in the act of landing her plane. We don't know if she and her copilot were observing the FAA's Sterile Cockpit Rule or were chatting about sports. It doesn't matter. The instant the stick shaker went off nothing dramatic changed for this captain. The rattling yoke entered seamlessly into her knowledge of her *situation* at that instant, and she reacted appropriately.

Like Renslow and Shaw one month earlier, both she and her copilot never noticed the INCR REF SPEEDS alert on their engine instrument display, nor did they notice the *low-speed cue* on their airspeed indicators. But she was aware of everything else about her *situation*, including her altitude, her plane's precise position on the final approach course to the airport, what the next few seconds of her flight would involve, her plane's engine power settings, and an infinite number of other data points she had been consciously and unconsciously observing and gathering.

Not so for Renslow. For him, the noisy stick shaker thrust him violently from one mode – chatting and monitoring, albeit not very closely – into another – reacting to an emergency. He was not dialed into his plane's *situation* the way the Vermont captain was, because his autopilot was doing its usual perfect job of bringing them down the glide slope.

He probably had not been closely following his flight's progress for a while. The need to slow his plane down from 184 knots at that late phase in their approach to Buffalo was testament to his failure to stay aware of his plane's precise *situation*. It is good he eventually realized they were going too fast, but he should have noticed that earlier. He wasn't engaged. He wasn't in the loop. He wasn't focused on the landing.

The same goes for Shaw. She too was not nearly as in tune with her plane's *situation* as she should have been.

I have a reason for repeating and italicizing the word *situation* in the preceding paragraphs. The Vermont captain's ability to know her *situation*, and Renslow's and Shaw's lack of knowledge of their *situation*, incorporates one-half of one of the two most important words in aviation human factors: *Situational Awareness*. Like most things in the flying business, its acronym, SA, is used as often as the words.

SA means knowing everything about your plane and your flight: your altitude and if it is changing, your airspeed and if that is changing, what you are flying over that very instant and what you will be flying over in a moment or two, your heading now and your heading in a few minutes, your final destination and how you are planning to get there, how your engines are performing, your plane's fuel level and how fast it is depleting, other planes in the sky that may be near you and are they a threat, and on and on. *Everything.*

Putting it in earthbound context, say you are driving on a busy highway to Aunt Sally's house for Thanksgiving dinner. As you're motoring along, if you are *aware* that ... the exit to her house is three miles ahead, you are driving at the speed limit, you'd just moved into the second-to-the-left lane because a car behind you was gaining fast and you wanted to let it by, you will arrive five minutes early, you have enough gas to get to her house and home again, and even though it is cloudy, rain won't begin until tomorrow morning ... then your *situational awareness* is excellent.

But now suppose you are on the same trip to Aunt Sally's, only this time you're thinking about that story she tells every year about her seven cats, you are worrying about the project your boss wants completed by next Tuesday, and you didn't sleep well last night. Suddenly the exit to her house looms up and you're in the wrong lane to make it off the highway without slicing recklessly across traffic. Here your *situational awareness* is poor.

Whether you are flying your single-engine Cessna forty miles to meet a friend for a Saturday morning breakfast or you are in command of a 747-400 flying 450 passengers 5,000 miles across the Pacific Ocean, there is no substitute for situational awareness and no excuse for not having it. And you either have SA or you don't. You cannot sort-of have it. You cannot be mostly aware of what's going on. If you're not 100% situationally aware, then you aren't aware at all.

Without situational awareness it is impossible to *stay ahead of the airplane*, as pilots describe keeping abreast of all the things needing doing in a cockpit. Pilots have to navigate, lower or raise landing gear, flaps and slats, monitor their airspeed and altitude, change air traffic control frequencies and talk to them, change navigation frequencies and make sure they are working, change headings at the right time, and watch for other airplanes. And that is an abbreviated list. When pilots *fall behind their plane*, they rush, miss radio calls meant for them, miss turns along their route, don't properly hold their altitude and airspeed, and make other mistakes. Mistakes in the air threaten lives.

With situational awareness, pilots plan what needs doing, they do them on time, and they stay ahead of their plane.

SA is often confused with *Spatial Orientation* or its opposite number, *Spatial Disorientation*. They are not the same. Spatial orientation and disorientation refer to knowing where you are in space, or as I put it more succinctly a few chapters ago, knowing which way is up. You can lose situational awareness while still being spatially oriented, while still knowing where the ground is. But if you are experiencing spatial disorientation you've also lost situational awareness.

According to some, German First World War flying ace Oswald Boelke first introduced the concept of situational awareness, though not in those exact words, as a means of gaining advantage on the enemy in aerial combat. Considered the father of air-to-air fighter tactics, as a combat pilot Boelke knew that being better *aware* of your overall *situation* than your enemy gave you an edge as you maneuvered against him seeking tactical advantage in 3-dimensional space.

But in civilian flying, situational awareness was never a uniquely defined goal pilots consciously sought to achieve. Instead, they did all the things now included within SA, but piecemeal – they stayed ahead of their airplane, they knew the weather, they knew their altitude and airspeed, and they knew where they were and where they were going. Combat pilots came closer to having true SA, adding to their situational knowledge the location of other planes in the sky, whether they were friend or foe, where they were going, how they were armed, and their friendly or hostile intentions.

No matter who gets credit, the actual word-pair *situational awareness* came about in the 1980s, providing a single basket to put everything related to being 'in the loop.' This is also roughly when human factors experts began contemplating the damage automation could do to a pilot's overall awareness, launching a field of research that hasn't stopped growing.

Automation used properly helps with SA. With the autopilot doing the flying a pilot can spend a minute studying a map to plan a route around thunderstorms cropping up ahead, or trouble-shoot balky landing gear. Automation's data collation and dissemination ability ideally keeps pilots perfectly informed of the condition of their plane. With the Flight Management System's computers properly programmed, a commercial jet can fly complicated routes in and out of busy airports accurately and safely, with the pilots watching for other planes and keeping an eye on their instruments.

And that is the catch, and maybe even the Catch-22. If pilots are only monitoring their instruments and looking out for other airplane traffic, how are they maintaining complete situational awareness? Automation, designed to improve situational awareness, can actually be a serious threat to it. It takes pilots' hands off the controls and demotes them from first-string players into benchwarmers, subs behind the all-electronic starting line-up.

While hand-flying, when the crew of a jet sees – either by looking out of their cockpit window or by using their navigation displays – that they have reached the point where a 30° left turn is called for, as they turn their yoke or tip their sidestick to bank their plane they know exactly where they are.

But when flying on autopilot, if the pilots know that when they reach a certain radio-beam-defined waypoint the FMS will make the turn for them, they don't need to concentrate on it. It will happen without them. So how attentive should they be if something they implicitly trust is controlling their route? If, after thousands of hours in the air, a captain knows the FMS has made every turn every time, how concerned will that captain be about the next one? Not very. Especially if he is distracted by other issues, like turbulence along the route or a sick passenger in the cabin.

We can now appreciate more completely, and more accurately identify, the impact automation had on Renslow and Shaw: they lost situational awareness. They lost it because, to use a word I introduced in the last chapter, they were *complacent*.

Indeed, they were victims of a demon facing every pilot since Lawrence Sperry invented the gyrostabilizer: *Automation Complacency*.

Automation complacency is Enemy Number 1 on the list of every human factors researcher, whether they are studying pilots, locomotive engineers, or factory managers. It leads directly to pilots losing situational awareness and it has become a major topic of discussion in flight crew rooms, airline boardrooms and academia.

I have found a host of definitions for automation complacency, some in plain language, others in psycho-babble, but parsing the term, looking carefully at the individual words, gets us to the heart of it. The definition for automation has two distinct elements: *what it is*, and *what it does*. I found both best described in a research piece co-authored by Raja Parasuraman, one of the leading lights of the human factors field.

Born in New Delhi in 1950 and with an electrical engineering degree from the University of London, Parasuraman became an expert in a human factors sub-specialty whose name he coined: *neuroergonomics*, the study of the brain mechanisms underlying human behavior and performance. He focused intensely on man-machine interactions, and is considered one of the world's great thinkers on the subject. This, while becoming a gourmet cook on the side.

In 1999 Parasuraman, at Catholic University of America, along with Thomas Sheridan of MIT and Christopher Wickens of the University of Illinois, defined automation (*what it is…*) as 'a device or system that accomplishes (partially or fully) a function that was previously, or conceivably could be, carried out (partially or fully) by a human operator.'

Your toaster-oven that turns itself off after 120 seconds is automated. So are your car's headlights that turn themselves on when day turns to dusk.

In the same piece Parasuraman and his co-authors describe automation's function (*what it does…*) as along a continuum, a range of machine-human exchanges. It begins with automation providing a human with a series of alternative ways to do a specific task, though not doing the task itself. It goes next to automation narrowing down the alternatives to a best-choice for doing the task, though again not doing it for the human. Then it goes to automation doing the task and informing the human, and ends with automation doing the task and not telling the human.

Complacency is trickier, with definitions from Merriam-Webster's through published psychological research having the complacent person experiencing some form of 'smugness,' or 'self-satisfaction.' Not to argue with the dictionary, but those definitions don't match the real world when assessing aviation accidents. Removing the arrogance component yields a much simpler definition – my own: *losing focus or concentration because of a firm belief – conscious or even unconscious – that all is good and right with whatever is being watched.*

To me, complacency is not about the superior or righteous attitude of the human observer. Rather, it is about the observer's steadfast belief in the superiority and righteousness of the automation.

Putting the words back together, automation complacency is what happens when a human monitor is watching an automated machine, though not very closely, because the monitor is absolutely certain – the monitor *unfailingly trusts* – the automation will keep doing its job.

THE DANGERS OF AUTOMATION IN AIRLINERS

No doubt automation complacency was a problem since the first days of the Industrial Revolution. In the 1700s, mechanical looms and cotton gins began taking the place of human-power for repetitive tasks like weaving textiles and separating cotton fibers from seeds. Watching these machines required effort to remain alert and avoid becoming complacent. Waning concentration would lead to jams and work stoppages. But the repercussions of a complacency-driven mistake were not very high – some lost production time, perhaps some damaged fibers.

For aircraft and their pilots it is a completely different story.

Commercial pilots today spend a few minutes taxiing their plane to the runway, hand-fly the takeoff, and then within a few seconds or minutes of leaving the ground they engage the autopilot and autothrottle, which stay on until some point during the approach to the destination runway. While automation is in control, the pilots have one primary task – monitoring the plane's computers and systems. There are exceptions – altitude and routing changes for instance, and they stay in touch with Air Traffic Control. But other than that, pilots are paid to watch and not touch. In a six hour coast-to-coast flight they may watch for five hours and fifty five minutes, and hand-fly for five.

Automation complacency can sneak up on pilots without warning. When it does, they can lose situational awareness without realizing it. Most of the time – really, nearly all the time – loss of situational awareness leads to nothing more than a few frenzied moments in the cockpit as the crew works to get it back.

But when an emergency strikes, when a stick shaker starts shaking or warning lights illuminate, loss of situational awareness can be a killer.

Chapter 18

Hazardous States

Lawrence Sperry's unplanned co-ed swim in Great South Bay with Dorothy Rice Peirce is a classic example of the dangers of automation complacency. Sperry and Peirce expected the Sperry Automatic Pilot to keep their biplane straight and level while they were engaged in, umm … non-flying activities. Whether one of them kicked the autopilot with their bare foot, or something else went wrong, they were so confident the device would work as designed that they failed to notice their plane was in trouble. They were complacent, and they nearly died because of it.

Because the consequences of complacency by pilots are so dire, psychologists working to understand the human factors phenomenon of automation complacency have focused much of their efforts specifically on aviation. And because no aspect of aviation is riskier than spaceflight, NASA, the National Aeronautics and Space Administration, maintains a robust practice in the psychology of automation at its Langley Research Center in Hampton, Virginia. It has to, because astronauts in orbit have none of the advantages available to those monitoring automated systems closer to the ground. The engineer monitoring auto factory welding robots can stop the line and take as much time as necessary to puzzle out a problem. Airplane pilots have as much time as their altitude and airspeed will allow them. In outer space, time is an even more precious commodity, and astronauts have no safe havens.

In the 1980s Tennessee born and bred Alan Pope, a scientist at NASA's Langley Research Center, began thinking hard about various states of human consciousness and how they related to pilots in their cockpit work environments. With a master's degree in electrical engineering, a doctorate in clinical psychology, and experience acting and singing in community theater, Pope had the perfect background for it.

He began by searching through and cataloguing responses to a self-reporting database managed by NASA called the Aviation Safety Reporting System, ASRS in the obligatory acronym. Pilots use the ASRS to confidentially self-report, in their own words, both minor and major incidents in the air. Pope's efforts led him to a conclusion.

'I discovered,' he said, 'that pilots sometimes provided on ASRS descriptions of their experiences of what we termed hazardous states and we went from there to hypothesize factors that contribute to them.'

His research culminated in 1992 in a report, co-authored with Edward Bogart, also from Langley, suggesting we can occasionally experience forms of consciousness – they called them *conscious states* – that may be dangerous to ourselves and to others because we are putting them at risk.[25]

Most of the day we are in our 'normal' conscious state, awake and taking in our environment through our five senses. We routinely take in more or less, depending on what we are doing – walking, talking, texting, working, driving, playing sports, eating dinner, reading a book, watching a movie, or whatever it might be that has our attention.

Then there are those states when either we are less vigilant than we are *supposed* to be, or on the flip side, when we are excessively absorbed with a thin slice of what we are watching, to the exclusion of everything else. That is known as *tunneling*. Pope and Bogart called all these conscious states 'Hazardous States of Awareness,' or HSAs.

In very unscientific language, a 'less vigilant' HSA is when we are not *all there*, when we are not *in the moment*, not completely *present*. Daydreaming, something we all do regularly, is an example. An *Excessive Absorption* HSA is when we are intensely focused on a single point, a single screen, a single output source, or preoccupied with something not even in the room while ignoring everything else around us. We are seeing the world through a drinking straw. Focusing exclusively on our car's gas gauge flirting with the 'E' while driving past our highway exit is an example.

Airplane accidents caused by aircrews caught in a Hazardous State of Awareness are usually the result of a less vigilant HSA. Pope and Bogart call them *diminished alertness* or *diminished awareness* states. Being excessively absorbed and drinking-straw-focused causes accidents as well, though less frequently. Sometimes they both occur in the same incident, one after the other: an excessively absorbed state could be the end result of a diminished awareness state having gotten the pilot into trouble. To work himself out of the jam he overcompensates by focusing too much on what he had ignored earlier, and that merely compounds the problem.

---

25 Bogart was employed by Lockheed Martin but worked as a contractor at Langley.

Being in an HSA is not dangerous in the right setting. If you are in a classroom struggling to stay awake as the professor drones on, or at the dining room table barely listening as Aunt Sally relates her cat story, you're slipping into an HSA that won't hurt you. You might even welcome luxuriating in that *diminished awareness state* for as long as you possibly can. But when your job is to monitor a system, if you are not one hundred percent engaged and locked in (though not *too* locked in), you are in an HSA, and that *could* be dangerous. For pilots, it could cause a loss of situational awareness, with all its unwelcome consequences.

Ten years after Pope and Bogart published their work, Lawrence Prinzel, also from NASA Langley, came up with a number of separate but related HSA *constructs*, or labels for identifying and categorizing HSAs.[26]

When we daydream, for instance, we are here but our mind is elsewhere. Daydreaming falls under the HSA construct Task-Unrelated Thoughts. While you are piloting your plane, if you are also vividly reliving last week's pick-up basketball game, for obvious reasons that is a Task-Unrelated Thought. Daydreaming also appears in a second construct, Boredom. If you are bored, you are likely to daydream.

Lapses and Slips comprise a third HSA construct. A Lapse is an error of omission: you walked into a room to get something, but as soon as you entered you forgot why you were there. Or you opened the refrigerator door and immediately forgot what you wanted. A Slip is an error of commission: you moved offices from the 23rd to the 27th floor, but your first day in the new location you got off the elevator on 23.

A fourth construct is Mental Fatigue. If in your past you have worked at your job very hard or driven an automobile very far, you have experienced some version of it. It is not clear if Mental Fatigue is purely physical, or if it also has a psychological component. If at work or school you have been exhausted and suddenly found powers of concentration to rally and complete your project or homework assignment, you would be uncertain if your fatigue was mental, physical, or some combination.

Prinzel identifies Blocks as a fifth HSA construct. He is alluding to mental blocks, when no matter how hard we try we just cannot come up with that word (and sometimes the harder we try, the more elusive the word). A pilot should never have a Block, never say, 'What's the name of that

---

26 A 'construct' in psychology is not actually a 'label,' but it serves our purpose in understanding HSAs and their impact on aviation safety to think of them in that simple way. I paraphrased the *labels* definition from the definition of Construct (Psychology) in Brittanica.com.

switch I'm supposed to press now?' because it should be on a checklist. Failure to use a checklist is not a Block, or even a Lapse or Slip. It is an inexcusable error in procedure.

Finally there is the construct we have already encountered: Complacency. Prinzel says this sixth construct 'represents a failure to monitor the actions of a machine or computer…under habitual and familiar circumstances.' That is even broader than my definition of *Complacency* a few pages back. Prinzel doesn't even require trust. All he requires is familiarity, which can come from observing the same thing over the past few years, or even just the past few minutes.

Prinzel doesn't say as much, but all six HSA constructs have within them an element of Complacency. We were complacent when we Slipped by stepping off the elevator on the wrong floor, because it was our old habit. We were Bored, so we grew complacent and began daydreaming and having Task-Unrelated Thoughts. We were complacent so we Lapsed and forgot to set the alarm before leaving the house. We were Excessively Absorbed by our gas gauge and became complacent about watching for our exit.

Everything circles back to Complacency.

Renslow and Shaw on Colgan 3407 can be described in terms of Hazardous States of Awareness. They were Fatigued (physically definitely, mentally maybe), and they experienced a Lapse by forgetting about the **REF SPEEDS** switch. They were on an approach to Buffalo airport that seemed routine to them – Niagara International itself may not have been familiar (the NTSB report doesn't indicate how often either of them had flown into Buffalo), but the drill of setting the autopilot to fly an approach down to a runway was sufficiently familiar for them to be conversing inappropriately below 10,000 feet. So perhaps they were experiencing Boredom, while their conversation qualifies as Task-Unrelated Thoughts.

And in each of their HSAs, Complacency played a role.

Chapter 19

Engine Gauges

The year after Pope and Bogart produced their work on Hazardous States of Awareness, groundbreaking research led by Raja Parasuraman honed in on the ultimate cause of complacency in the cockpit. In 1993 Parasuraman joined with two associates, Robert Molloy and Indramani Singh, to research what sort of cockpit environment would most likely bring about complacency. Specifically, they wondered if a pilot would be more likely to become complacent if things were humming along smoothly, or if alarms and warnings were going off with some frequency.

Turns out the answer is not obvious. The report Parasuraman and his partners wrote describing their research and conclusions is one of the most important ever produced on the subject.

To get their not-obvious answer, Parasuraman, Molloy and Singh constructed an experiment around a unique personal computer screen dubbed a Multi-Attribute Task Battery – MATB. Designed originally by Ray Comstock at NASA's Langley Research Center, the MATB display screen is divided into three discrete sections. One roughly mimics flying a plane on a flight, the second manages fuel for that flight, and the third displays information on the plane's engines.

Test subjects spent two hours 'piloting' in front of a MATB screen by using a joystick to line up cross-hairs moving along a track. Like real pilots in a real cockpit, at the same time the test subjects needed to pay attention to their fuel supply. Fuel pumps required turning on and off once in a while to keep fuel flowing. While the subjects were busy 'flying' along the track and managing their fuel, they also had to keep an eye on the third screen section, the one showing engine data.

The engine section was the heart of the experiment – where the trouble would be. It contained four engine gauges: one recording temperature and one monitoring pressure for each of the plane's two engines. The section also contained two lights above the cluster of gauges, green and red.

Green signified all was good with engine temperatures and pressures. It was lit most of the time.

Red indicated that one of the four gauges had gone out of whack. Within four seconds the computer would automatically detect which gauge had the problem, and then fix it. The gauge would return to normal, the red light would blink off, and the green light would illuminate again.

That is what was *supposed* to happen.

Here was the experiment: once in a while one of the four gauges would reveal an engine problem, but the green light would stay on and the computer would *not* fix it. For instance, a gauge might show low pressure in Engine #2, yet the green light would remain lit and the red one dark. In this case the test subjects were expected to notice the errant gauge on their own, and press a specific keyboard key to 'fix' the problem themselves.

The question was: how well could the subjects fly the plane and manage their fuel while simultaneously monitoring the engine gauges, ready to pounce if the automation failed to detect a misbehaving engine? How frequently would they miss a bad gauge reading?

Parasuraman's experiment went one step further. He inserted a variable, part of the experiment that would change among the test subjects. That variable was how often the computer failed to detect and fix an engine problem. Put another way, the reliability of the computer's detection capability would differ among the test subjects.

The subjects were split into two groups: *constant-reliability* and *variable-reliability*.

To Parasuraman, *constant reliability* did not mean the computer would consistently detect and fix each trouble incident within the allotted four seconds. It meant the computer was consistent in its *trustworthiness*. He sliced this *constant-reliability* group into two sub-groups: one subgroup could trust the computer to almost always detect and fix the engine trouble (actually 86.5% of the time). The other could trust the computer to miss the engine trouble nearly half the time (detecting and fixing it 56.5% of the time).

Parasuraman's computer was like your teenage nephew who had promised to come by at 4:00 every afternoon to walk your dog. For the *constant-reliability* group, one subgroup knew he was highly trustworthy, at their doorstep right on time 86.5% of the time. The other subgroup knew half the time he would be an hour late. For them, they would be thinking, 'I can trust him to forget nearly half the time.'

The *variable-reliability* test subjects didn't know what to expect from the computer. That is, they had no idea how trustworthy it would be.

Sometimes it would detect and fix engine trouble at the trustworthy rate of 86.5%. Other times it would be lax, detecting and fixing problems at the untrustworthy rate of 56.5%. It was like your teenage nephew being on time some days and an hour late on others. It is nice that he would always show up, but in truth he was untrustworthy, completely unreliable.

When the test concluded, Parasuraman and his team discovered both *constant-reliability* subgroups were meaningfully worse at monitoring the engines than the *variable-reliability* group. That led him to his big conclusion: the key to high-quality, vigilant monitoring was the trustworthiness of the automation doing the monitoring. When the automation could be trusted to be either consistently great or consistently poor, the test subjects were 'lulled into complacency,' to use the words in Parasuraman's report. In other words, if you knew your automation would either catch errors every time, or hardly catch them at all, you would become complacent.

But when the automation kept subjects guessing, when its automatic problem-detection performance was inconsistent and untrustworthy, then monitoring vigilance was much better. Back to your nephew, if you knew he would always be on time or always be late, you'd know how to handle it – and you would become complacent. But if some days he was perfect and other days not so much, you would be on your guard, not knowing how to plan your afternoon – you would be more vigilant.

One part of this result is not intuitive to me. I would have expected a highly-unreliable computer would lead to increased vigilance. Yet that was not one of the results. Parasuraman shared this concern, writing, 'It is possible the "low" value of reliability [56.5%] used in the present study was not low enough to offset the complacency.' He was suggesting that if he had jacked up the failure rate, complacency might not have set in.

That feels right to me, but it doesn't take anything away from the main conclusion of his experiment: *high reliability, a strong trust in technology, leads to complacency*. Only if we don't trust the machine, if we don't feel a familiarity with it, does vigilance stay high.

Renslow and Shaw on Colgan 3407 trusted their autopilot, and that led to Complacency, a Hazardous State of Awareness.

Complacency caused them to lose situational awareness.

Because they lost situational awareness, when the stick shaker shocked them back to reality and the autopilot-disengaged horn went off, Renslow panicked, and Shaw was transported back to her flight school days.

Part III

Landing Gear Down

Chapter 20

San Francisco

San Francisco International Airport, widely known simply by its aviation-industry identifier, SFO, is tucked into the western shore of San Francisco Bay thirteen miles south of the famous Golden Gate Bridge. On clear days pilots flying in and out are rewarded with spectacular views of the scenic Bay Area. Saturday, 6 July 2013, was going to be one of those days. Rising just before 6:00 am, the sun revealed a few scattered clouds as San Francisco woke to typically cool morning temperatures in the 50s. By noon it would be 72 degrees.

High over the Pacific Ocean, Asiana flight 214, a Boeing 777-200ER (for Extended Range), was still in darkness half way through its eleven-hour flight from Seoul, South Korea, to SFO. In its hushed passenger cabin were 291 mostly sleeping passengers, 12 flight attendants, and 4 cockpit crewmembers.

Four pilots was a typical number for Asiana's long-haul transpacific flights: a primary crew of captain and copilot, and a two-person relief crew so the primary crew could take a mid-flight break. But their makeup this day was unusually top-heavy. Occupying the left seat, the captain's position, was 45-year-old Lee Kang-kook. Highly experienced with 9,684 hours in his logbook, all of it as an Asiana pilot, for the past six years he had captained Airbus A320s. He was now transitioning to the captain's seat in the larger Boeing 777 and was nearly done with his training.

As part of the last phase of that training, Kang-kook was being observed and graded on this flight by Lee Jeong-min. The 49-year-old senior captain had just been promoted to Instructor Pilot and was sitting in the right seat.[27] This would be his first flight as an instructor. Starting his flying career in the Republic of Korea Air Force, he joined Asiana in 1996 and now had over 12,000 total hours in the air, one-quarter of that in the Boeing 777.

---

27 Korean names are given by family name first. As two of Asiana 214's pilots have the same family name, *Lee* (they are not related), I will refer to the pilots by their first names to avoid confusion.

79

THE DANGERS OF AUTOMATION IN AIRLINERS

As the sun rose over the Pacific that morning, the relief aircrew was in charge. Then at 9:38 am (all times for Asiana 214 are in San Francisco's time zone, Pacific Time), following five hours resting in the crew-rest area of the passenger cabin Kang-kook returned to the cockpit and took his seat on the left. Ten minutes later Jeong-min returned to his seat on the right. For the approach and landing Kang-kook would be the Pilot Flying while Jeong-min would be the Pilot Not Flying, working the radios and monitoring the flight, as well as critiquing Kang-kook's performance.

An hour later, and as part of their standard operating procedure before descending, the CVR recorded Kang-kook briefing Jeong-min on approach and landing details, including mentioning his expectation of being directed by air traffic control to fly a *visual approach* for Runway 28-Left. SFO's parallel runways 28-Left and 28-Right are approached from San Francisco Bay and their thresholds are right at the water's edge.

With great weather at SFO, it was not surprising that Kang-kook's briefing included a *visual approach* instead of an instrument approach. But SFO was using that approach not only because visibility was nearly perfect, but also because the runway's ILS glide slope beam was not working. Normally that should not have been a problem. Most pilots would relish the chance to bring their jet down using just skill and judgment.

But not Kang-kook. He would have preferred the help of the runway's ILS glide slope radio beam to double-check his flight path. Its unavailability was stressing him, but he kept that to himself.[28]

For student pilots, learning to visually judge a plane's descent path to the runway isn't easy, and it comes more quickly to some than to others. They do not get their first level pilot license until they have conquered it. Then it takes more practice learning to manage that path in a big, heavy commercial passenger jet. Using the cockpit instrument that tracks an ILS glide slope radio beam makes the process easier: the pilot just needs to stay on the track shown on the cockpit screen, like it is a video game.

Without a doubt, with nearly 10,000 hours in the air Kang-kook could judge a *visual approach* to a successful landing without help. But he had never done it in an actual Boeing 777 with passengers behind him, only in a simulator. While the simulator was full-motion and as realistic as technology could make it, if there was ever a good day to do it for real, this was it.

---

28 The NTSB report on the crash uses the word 'stressed,' in quotation marks, implying it was either the word Kang-kook himself used in a post-accident interview, or a judgement of his mental state and attitude by NTSB investigators.

Besides, he didn't have a choice.

The international language of aviation is English, but within a cockpit pilots tend to chat in a mixture of English and their native tongues. After Kang-kook's briefing, switching between Korean and English the two pilots discussed a few additional landing details, then drifted into light banter. They covered sunglasses, both agreeing it hurt depth perception so they don't wear them during landing. They discussed kids with bad eyesight, flight physicals, and discount passes for crewmembers offered by Asiana and other airlines. Like Renslow and Shaw, they weren't all business while cruising at altitude. But unlike the Colgan 3407 flight crew, once they descended below 10,000 feet the Sterile Cockpit Rule went into effect and their chatter ceased completely.

At 10:55 that morning, air traffic control called the flight. 'Asiana two fourteen descend pilots discretion maintain flight level two four zero.'

It was time to start coming down. They had permission to descend to 24,000 feet *at their discretion*, whenever they were ready.

One minute later they were all set. As the Pilot Not Flying, Instructor Pilot Jeong-min keyed his mike and said, 'Asiana two one four leaving three nine zero for two four zero.' They began their descent from 39,000 feet.

At 11:12 the relief copilot, 40-year-old Bong Dong-won, with 4,557 hours of total flight time and another former Republic of Korea Air Force officer, returned to the cockpit. Sitting in the fold-down jump seat behind and between the two captains, he had a clear view out of the cockpit windows and of the entire instrument panel, including both pilots' primary flight displays. Not obligated to be there, he knew he could always learn something new watching an experienced crew expertly doing their thing.

Jeong-min acknowledged his presence, saying to him in Korean, 'monitoring well please in the back ... let us know immediately if anything strange shows.'[29]

The relief captain remained in the passenger cabin. Not needed on the flight deck, he would stay in the back for the duration of the flight.

A few minutes later and now under the control of Northern California Terminal Radar Approach Control, *NorCal Approach*, Asiana 214 flew directly over SFO at 11,000 feet. NorCal then guided Kang-kook through a series of straight legs and turns directing him towards Runway 28-Left's *final approach*.

---

29 This verbatim quote from the NTSB report on the crash is an imperfect translation from Korean.

At 11:22 am NorCal radioed, 'Asiana two one four heavy ... cleared visual approach runway two eight left.'[30]

NorCal was permitting Kang-kook to fly to the runway, clearing him specifically for the approach he was 'stressed' about, the *visual approach*. The controller had placed him right where he needed to be, on the localizer beam for the runway, and on the glide slope beam if the glide slope had been working. Jeong-min confirmed to NorCal that he heard and understood their instructions, and in the cockpit Kang-kook acknowledged the same to his crewmates.

In spite of his worries, Kang-kook was doing a solid job flying, as would be expected of a captain of his long experience. He had taken them down from 39,000 feet and cruising speed to now 5,300 feet and 210 knots and nearly lined up with Runway 28-Left as he continued descending.

Though Kang-kook was the Pilot Flying, like Renslow in Colgan 3407 he was not hand-flying. He was controlling the Boeing through its autopilot and autothrottles, flying by adjusting knobs on a piece of instrument panel real estate called the MCP, the Mode Control Panel.

The MCP is a strip of buttons, knobs and small display windows just below the glareshield between the two pilots and comfortably reached by both of them. By turning and pushing the MCP's knobs and buttons Kang-kook could change the plane's airspeed, altitude, and heading, as well as the autopilot and autothrottle *modes* dictating how each would behave. In some ways it was merely an updated form of the sort of aircraft control performed by bombardiers in the 1940s.

When NorCal had given him permission to descend to 24,000 feet, Kang-kook had twirled a knob on the MCP to put **24000** in the appropriate window, and then the autopilot had done the rest. After NorCal had asked him to slow the plane to 210 knots, he had twisted another knob on the MCP to **210**, and the autopilot had handled that too. And when NorCal had assigned him new headings, he had dialed in those headings and the autopilot had banked the jet into the turns.

Now that Asiana 214 was just about lined up with the runway, Kang-kook pressed a button on the MCP to set the autopilot mode to **LOC**, *Localizer Mode*, locking on to Runway 28-Left's localizer radio beam. The autopilot would keep his plane pointed at the runway centerline while he handled the details of the descent.

He began flying the final approach.

---

30 The designation 'heavy' is appended to the flight number of the largest aircraft, specifically planes with a maximum takeoff weight in excess of 300,000 lbs. Crews don't often use it, but air traffic control generally does.

Chapter 21

Line in the Sky

A well-known axiom in flying holds that a good approach leads to a good landing. Slipping a plane into a smooth and steady descent, painting a straight line in the sky with just a few tiny tweaks of the throttle or elevator over the final few miles to the runway is called a *stabilized approach*, and is likely to conclude with a sweet chirp as the tires kiss the concrete.

To produce a stabilized approach, to draw that straight line to the runway, pilots need to get their plane descending at just the right downward speed, the right *rate-of-descent*. [FIGURE 14] That line exactly overlays the radio beam producing the invisible glide slope on the runway's ILS approach. The correct rate-of-descent depends on a plane's airspeed: the higher the airspeed, the faster a plane must descend to stay on the beam. The *vertical speed indicator* on each pilots' primary flight display shows their rate-of-descent in *feet per minute* (it also displays the rate-of-climb when a plane is going up).

Taking the guesswork out of the approach, Asiana's cockpit crew had access to SFO *approach charts* listing the precise rate-of-descent for their airspeed to keep them perfectly on the glide slope line.[31] The chart showed that at their airspeed during this early stage of the final approach Asiana 214 needed to descend at 1,000 feet per minute. Further along the approach, as they neared the runway and slowed down further, Kang-kook would need to reduce the rate-of-descent to 700 feet per minute. Just to compare, if he had been flying a small single-engine Cessna with an approach speed less than half the 777's, he would have used a descent rate of 450 feet per minute.

---

31 *Approach charts* detail everything a pilot needs to know for each approach to every runway at an airport. In the past they were in paper form and pilots kept them in loose-leaf binders, carrying them wherever they went and updating them monthly. Today they are digital, displayed on a pilot's tablet or on one of the cockpit screens, and updated electronically.

Asiana 214 was fifteen miles from Runway 28-Left and descending at 900 feet per minute. They should have been descending at 1,000 feet per minute, but it was early in the approach. They had a little time.

Fifteen miles is a long way to accurately visually judge a descent, so to make the approach more manageable Kang-kook first aimed at a waypoint called DUYET nine miles directly in front of them.[32] This was like a golfer facing a long putt aiming at a discolored blade of grass half-way to the hole. Kang-kook could not see DUYET looking out of the cockpit window – it was an electronic intersection in the sky. But the SFO approach chart showed if Kang-kook was at 1,800 feet when he reached DUYET he would be right on the glide slope.

He ordered the Boeing to descend to 1,800 feet by dialing **1800** into the MCP. He also slowed the plane from 210 to 192 knots by twisting **192** into the MCP.

As soon as he entered **192**, NorCal radioed Asiana 214 and instructed Kang-kook to slow down more, to 180 knots, and to maintain that speed until five miles from the runway (half a mile past DUYET). Once he was within five miles he could slow further, to landing speed. To comply, Kang-kook dialed **180** into the MCP. He also asked Jeong-min to set 5° of flaps, because as the plane slowed the wings needed the increased *lift* from the greater camber line the flaps provided.

The autopilot and autothrottle of a Boeing 777-200ER are a remarkably complex pair of avionics devices. The combined system is called the *Automatic Flight Control System*, or simply *autoflight*. Autopilot and autothrottle each have numerous operating *modes* dictating how they will react to every situation.

Pilots must have the details of every autoflight mode memorized, recallable instantly. Some modes are set by the pilot using buttons on the MCP. Others are set automatically, and changed automatically, by the plane's computers. Pilots need to know when and why modes will change by themselves. More than one accident we will read about was caused by pilots not noticing a mode had changed.

The multitude of mode combinations has made *mode confusion* an important human factors topic. It means exactly what it sounds like: cockpit crew confusion caused by not knowing what modes do, or how they interact with each other, or why they sometimes change on their own. The topic

---

32 Waypoints have 5-letter names. While they are all pronounceable, sometimes they are complete gibberish, other times they sound like something related to their location.

draws as much attention as automation complacency to the negatives of aviation automation.

When Kang-kook dialed **180** into the MCP, the Boeing 777's autoflight had two choices for slowing down. The obvious choice, the one we already know about, is reducing engine power – less power means less speed. The second choice is tilting up the elevator in the plane's tail, which would raise the nose and the angle-of-attack. The autopilot mode Asiana 214 was in reduced speed using this second choice – the elevator. Since understanding it is the key to understanding what followed, we need to take another brief detour. It will be our last.

Most people's familiarity with speed control comes from driving a car. Even non-drivers know a car has an accelerator pedal that causes its engine to rev more or less, making it go faster or slower. To go faster you depress the accelerator pedal more. To slow down, you lift your foot up off the pedal.

It is the same for an airplane. To go faster you add power by pushing the throttle forward. To go slower you pull the throttle back.

There is another way your car can speed up or slow down. If the road starts climbing a hill, your car will slow down on its own unless you depress the accelerator pedal a bit more. If the road starts descending a hill you'll speed up unless you let your foot off the pedal by a touch (or tap the brake).

This too is the same for an airplane. You can create your own hills in the sky by pulling or pushing the yoke, sidestick or stick. Pulling it back raises the elevator and makes your plane go 'uphill' and therefore slower. Pushing it forward lowers the elevator and makes your plane go 'downhill' and therefore faster.[33]

Autoflight using the elevator to control speed this way is called *speed-on-elevator*, since elevator movements are changing the plane's airspeed. By extension, autoflight using the throttle to speed up or slow down is called *speed-on-throttle*.

With one hand on the yoke or stick controlling the elevator, and the other hand on the throttle, pilots have infinite choices for climbing, descending and holding altitude while speeding up, slowing down, and holding airspeed. For instance, in Chapter 9 I said pilots can slow down without descending by reducing power and simultaneously tipping up the nose to raise the angle-of-attack. What I didn't say, but you can now safely assume,

---

33 Some planes also have brakes – they are called air brakes or speed brakes – that pilots sometimes use to slow down. They are sheets of metal that emerge from the wings or fuselage, creating extra drag.

is the reverse is also true: if you increase power while lowering the nose, you'll speed up but you won't climb.

When a pilot has been flying long enough, negotiating this dance between throttle/power and elevator becomes entirely intuitive. Or at least one hopes so.

Now back to Asiana 214 on the approach to SFO.

Kang-kook's autopilot was set to a mode called **FLCH SPD** – *Flight Level Change Speed*. This mode maintains whatever speed is in the display window by using *speed-on-elevator*. Autopilot tilts the elevator up to slow down (making the jet go 'uphill'), and down to speed up (making the jet go 'downhill'). So when Kang-kook dialed first **192** and then **180** into the MCP, the autopilot slowed down by raising the elevator. But raising the elevator without reducing power made Asiana 214 descend less quickly (instead of climbing). In fact it made the jet's rate-of-descent *much* less fast: it changed from 900 feet per minute to 300 feet per minute.

Not good.

Descending too slowly took Asiana 214 well above the glide slope. The plane was still far enough away from the runway to fix this error – and it was now a serious error – but someone in the cockpit needed to spot it quickly, ideally the Pilot Flying, Kang-kook.

Noticing the mistake should not have been difficult. Possibly Kang-kook could tell by looking out of the cockpit windows and using his judgment, though at this distance it might not have been very obvious. But he – and Jeong-min – also had a large cockpit display showing it graphically. Pilots on 777s each have two large flat-panel display screens in front of them. One is the PFD, the primary flight display. The other is the Navigation Display, pictorially showing Kang-kook and Jeong-min their final approach course down to the runway, with DUYET plainly marked. A green line on the screen showed *exactly* where on the final approach course their plane would reach 1,800 feet. The line was a couple of miles *past* DUYET, closer to the runway.

That was not in Kang-kook's plans.

The line would be closer to the runway only if they were descending too slowly. But it seems neither Kang-kook nor Jeong-min grasped that. Missing it confirmed Kang-kook was not giving this landing his complete attention. He may have already been in a Hazardous State of Awareness, perhaps Complacency, letting his trusty autopilot handle the approach.

Same for Jeong-min, who may have believed Kang-kook was experienced enough to handle a visual approach on a beautiful San Francisco day and so may have been both experiencing Complacency and having Task-Unrelated

Thoughts, daydreaming in the sun-drenched cockpit. The approach phase of his first flight as a newly-minted Instructor Pilot was not the time to be zoning out.

If the crew did not take corrective action quickly, the jet would land on the highway beyond Runway 28-Left.

It took more than fifteen seconds, a surprisingly long time at that airspeed, but eventually Jeong-min said something unintelligible, presumably regarding their altitude, and Kang-kook reacted. The first thing he did was slow down to 172 knots by changing the **180** in the MCP display to **172**.

That was a mistake – worse, it was two mistakes. We already saw the autopilot using *speed-on-elevator* when Kang-kook slowed the plane from 210 knots to 180 taking the jet above the glide slope. Slowing down more would exacerbate that error. Mistake Number One.

Mistake Number Two was Kang-kook did not have permission from NorCal to reduce speed. He was under orders to fly at 180 knots until they were five miles from the runway. He had already forgotten NorCal's instructions from just 35 seconds earlier. His head was not in the game.

To his credit, his next action was a step in the right direction. He pressed a button on the MCP changing the autopilot mode to **V/S**, *Vertical Speed* mode. In **V/S** mode he could drop the plane down to 1,800 feet quickly by dialing in any rate-of-descent he chose, and the autopilot would make that happen.

First he dialed in **-900**, down 900 feet per minute, which was roughly what the approach chart showed would be the right rate-of-descent to stay on the glide slope. But they had ground to make up, so that would not be enough. Then, perhaps remembering he was too high, he added 100 feet per minute, twisting in -**1000**, down 1,000 feet per minute.

Meanwhile, in response to setting the autopilot to **V/S** mode, the autothrottle switched itself automatically to **SPD** mode – *Speed* mode. It would keep the plane flying at whatever speed was set in the MCP. Right now that was 172 knots.

If all this sounds complicated and hard to follow, it is. It is automation at its finest. And they still had eleven and a half miles to go.

Switching autoflight to **V/S** and **SPD** modes was the right thing to do, though dropping at a rate-of-descent of 1,000 feet per minute was not enough. The cockpit Navigation Displays still showed that unmistakably – if only one of the crew would look at it. The green line was not where they needed it to be.

Thirty five seconds later Dong-won, the co-pilot in the jump seat, spotted the **172** knot airspeed error in the MCP. Nine and a half miles from

28-Left and at 3,900 feet altitude, he reminded them they were supposed to stay at 180 knots until five miles from the runway:

Dong-won: *To one eight zero five miles.*
Jeong-min: *Uhh … ah ah ah one eight zero*
Dong-won: *One eight zero five miles*
Kang-kook: *Huh?*
Dong-won: [something unintelligible, then…] *one eight zero*
Kang-kook: *Okay one eight zero five miles*

Jeong-min got it right away, but it had taken Dong-won three tries before Kang-kook became alert and reset the speed in the MCP to **180**.

And he was still too high.

Chapter 22

Cleared to Land

Pilots have three options if they are too high on final approach. They can descend faster, they can fly slower, or they can abandon the approach entirely and go around. Kang-kook was being observed and tested by Jeong-min, so the idea of giving up on this approach and going around for another try was highly unappealing. He would be failed for the flight, with probably serious repercussions for blowing a landing on a sunny Northern California morning.

As for his other two choices, Kang-kook could not fly slower – NorCal would not let him. But he could drop down faster. One of the ways to do that was by lowering the landing gear, adding drag to the plane. He ordered Jeong-min to bring down the gear.

As the Pilot Not Flying, Jeong-min pulled the landing gear lever. Then he alerted Kang-kook that in his view things were turning sour, saying in Korean, 'This seems a little high.' They were at 3,400 feet.

After two seconds Kang-kook said, also in Korean, 'Yeah.'

But he did nothing, which is difficult to understand. Maybe he was not only being complacent, he too was lost in a Task-Unrelated Thought. Or maybe he was mentally going in the other direction, becoming excessively-absorbed trying to figure out where he was messing up.

Hoping for a better reaction, Jeong-min repeated his concern, now saying in Korean, 'This should be a bit high.'

Another three seconds passed, a very long time flying at three miles-per-minute. Admittedly Jeong-min's last comment was not very clear. He could have been alluding to airspeed, altitude or rate-of-descent.

Kang-kook then responded, 'Do you mean it's too high?' He was referring to altitude.

Jeong-min replied with something unintelligible, perhaps a Korean 'Yes.'

Another three seconds elapsed before Kang-kook said, 'I will descend more,' and turned the knob on the MCP controlling their rate-of-descent to **-1500**. They were less than eight miles from the runway.

Dropping down at 1,500 feet per minute was an effective move, one that could get them back on the glide slope. It was a very fast rate-of-descent so close to the touchdown point, but this experienced crew should have been able to handle it. Maybe that is why Kang-kook had taken so much time before replying to Jeong-min – he was rummaging in his mind through his options. In spite of his past mistakes, things were looking up.

But not for long.

A half-minute later, six miles from the runway and at 2,600 feet, Kang-kook did something incomprehensible: he changed the rate-of-descent on the MCP back to **-1000**, back to 1,000 feet per minute. He had no reason on earth for doing that. Asiana 214 had not yet reached the glide slope, though it had been getting closer. Kang-kook categorically needed the 1,500 feet per minute descent rate if he had any hope of getting back on track.

While it is difficult to judge a plane's rate-of-descent fifteen miles from a runway, at six miles out – only two minutes travel time at their airspeed – they were close enough to definitively see the likely touchdown point, which right now was way beyond the far end of the runway. Did Kang-kook and Jeong-min realize they were so high? How about Dong-won in the jump seat, couldn't he see what was happening? Did any of them know how little time they had left to get things right?

Or had they lost situational awareness?

The Asiana pilots could have used a well-known landing rule-of-thumb, that a typical glide slope loses 300 feet altitude for each nautical mile travelled. They were six nautical miles from 28-Left, so 6 miles x 300 feet-per-mile = 1,800 feet. But Asiana 214 was 2,600 feet high, 800 feet higher than the rule-of-thumb dictated. If Asiana 214 continued descending from here at the rule-of-thumb rate-of-descent it would need nearly nine miles to reach the ground.

No one in the cockpit figured this out.

NorCal called the flight. 'Asiana two one four heavy contact San Francisco Tower one two zero point five.' They were being instructed to report in on the airport tower's frequency of 120.5 MHz (Megahertz).

Jeong-min radioed back to NorCal that he understood, but then turned his attention into the cockpit. Pilots are not required to immediately radio the tower when NorCal releases them, although it is not smart to dawdle either.

Now Kang-kook did something that seemed right at the time.

To 'stay ahead' of their plane during an approach, pilots plan what they will do if they have to abandon the approach and go around, executing a 'missed approach,' in the vernacular. A missed approach can happen for many reasons. If a plane that has just landed has not cleared the runway, the control tower will order the next plane in the approach sequence to go around. Or pilots may feel they are too high to make the runway, and choose on their own to go around and try again. Possibly Kang-kook was now considering he was too high to salvage the approach, and he would face whatever consequences that brought him.

When a plane goes around, the very first thing it does is climb away from the runway. To account for this possibility, Kang-kook dialed **3000** into the altitude display window of the MCP, telling the autopilot to take the plane up to 3,000 feet if Kang-kook issued the appropriate mode change. Since right now his plane was descending in **V/S** mode, vertical speed mode, the **3000** in the MCP was having no effect on anything.[34]

A few seconds later Asiana 214 crossed DUYET while flying at an altitude of 2,250 feet. Kang-kook had badly missed his goal of passing DUYET at 1,800 feet. The jet was still not descending quickly enough. A few more ticks of the clock, and they were five miles from the runway, the point where Kang-kook had permission to go slower than 180 knots.

Jeong-min tuned the radio to the tower frequency, 120.5 Mhz, and announced, 'Ah good morning Asiana two one four seven miles south two eight left.'

He received no response. The tower was busy handling other flights and would get back to them soon enough. Jeong-min had let the tower know he was listening on their frequency and that he was seven miles south of 28-Left. That was good enough.

Except that wasn't good at all. Asiana 214 was not seven miles from the runway – it was *five* miles away. The tower controllers had radar showing where the plane was and so the mistake was meaningless to them. They may not have even noticed. But how did Jeong-min make this mistake? He knew he had passed DUYET, which he *should have known* was five and a half miles from the runway. Why didn't Kang-kook or Dong-won correct him? It was more proof of his and the entire flight crew's lack of situational awareness while under automation's spell.

The plane flew on.

At 11:26 am they had four and a half miles to go, flying at 1,900 feet and 175 knots. Kang-kook asked his Instructor Pilot/Pilot-Not-Flying to

---

34 Because Kang-kook was flying a *visual approach*, he had no formal 'missed approach' instructions to work with, so he picked 3,000 feet as a random and safe altitude to fly up to if he abandoned the approach. On the other hand, when a plane flying an *instrument approach* goes around, its pilots must follow very specific *missed approach* routing instructions detailed in the runway's approach chart.

lower the flaps to 20°. Jeong-min complied. Soon after, Kang-kook twisted **152** into the MCP, to slow the jet to 152 knots.

That was good. He was slowing down on his way to their landing speed of 132 knots.

Fifteen seconds later, Kang-kook made his biggest mistake, the one that sealed their fate: he dialed **FLCH SPD** (*Flight Level Change Speed*) mode back into the MCP, replacing **V/S** mode. The change instructed the autopilot to fly to 3,000 feet, the go-around altitude Kang-kook had set into the MCP earlier. At the same time the autothrottle automatically changed itself to **THRUST** mode.

This is distinctly not what Kang-kook wanted. He had no excuse for this, no explanation whatsoever. Following his post-crash interviews with investigators, the NTSB report says 'he did not believe he had selected FLCH SPD.' Yet no one else could have done it, and autoflight didn't do it for him. He did it. And no one – not even he – knows why. His state of awareness was as hazardous as could be.

Meanwhile the big jet's autoflight complied with the new modes. Since they were still trying to slow to 152 knots, autoflight tilted the elevators up (*speed-on-elevator*), pitching the plane's nose upward. And since they were at 1,900 feet and now had to climb to 3,000, and because **THRUST** mode uses engine thrust to gain altitude, autothrottle moved the throttles forward, adding power.

Kang-kook saw the throttles advance out of the corner of his eye and knew immediately he had messed up. Moving fast, he quickly disconnected the autopilot by double-clicking a button on the yoke with his left hand, pulled the throttles back to idle with his right, and pushed the yoke forward so that the plane returned to descending at 1,000 feet per minute.

But he missed something.

Kang-kook thought that by pulling the throttle back he had merely stopped it from advancing. That was true. But his action also caused it to put itself – *by itself* – into a new mode: **HOLD** mode.

In **HOLD** mode the throttles sit at idle, no longer controlling the plane's engine power. They will do *nothing at all* until the pilot changes the autothrottle mode, or he reaches down and moves the throttles himself.

If Kang-kook had only looked at his primary flight display, he would have seen the word '**HOLD**' where the autothrottle mode was shown. And he should have known what it meant.

Jeong-min said, 'Speed,' reminding Kang-kook the plane was still flying too fast, so the Pilot Flying changed the MCP to **137**, five knots faster than landing speed, which was exactly the right speed for this part of the flight.

If you are thinking that Kang-kook's change to **137** made no sense because autothrottle was in **HOLD** mode, you're right. He did it because he did not realize autothrottle was not going to participate. Meanwhile Jeong-min lowered one final set of flaps, taking them to 30°, which would help slow the plane because of the increased drag.

The plane dropped lower.

Jeong-min said, 'It's high,' this time unambiguously meaning the plane's altitude was still too high, and in response Kang-kook now pushed the yoke further forward, plunging the jet down at rates-of-descent of between 1,500 and 1,800 feet per minute.

They were dropping so quickly that Dong-won blurted out 'sink rate sir' three times in nine seconds, alerting Kang-kook his rate-of-descent was excessive. This was definitely not a *stabilized approach*, the sort that produced a smooth line in the sky and a great landing.

At 11:27 they were only two miles from the runway and below 900 feet altitude. Less than one minute to go. Jeong-min needed permission to land. He radioed the SFO's tower again: 'Tower Asiana two one four short final.'

A tower controller replied. 'Asiana two one four heavy San Francisco tower runway two eight left cleared to land.'

They had permission. Now all they had to do was get there.

Twenty seconds later the plane was perfectly, and finally, on the glide path. Jeong-min announced to Kang-kook, 'On glide path sir.'

He and the other pilots knew this because they could see a bank of four bright PAPI lights just to the left of the runway. You can almost hear the relief in Jeong-min's voice.

PAPI lights, Precision Approach Path Indicator lights, change color between white and red, clueing pilots that they are high, low, or right on the glide slope to the runway.[35] The closer a plane comes to the threshold, the easier the lights are to see, and now all three pilots were watching the PAPI lights declare through colors that their plane was smack on the glide slope.

Kang-kook pulled back on the yoke to raise the elevator and arrest the rapid rate-of-descent, which was now 1,100 feet per minute. He needed the plane descending at 700 feet per minute for this last mile of the approach. Pulling back was the right move if autothrottle was working as he expected.

---

35 Most runways at major airports, and often at smaller airports has well, have PAPI lights. A slightly different version, called VASI – Visual Approach Slope Indicator – with the identical function, can also be found alongside runways.

But it was not working – it was in **HOLD** mode. Kang-kook still did not realize this, still had not looked at his PFD showing **HOLD** in the box for autothrottle mode. They were 380 feet above the waters of San Francisco Bay.

While everyone watched the PAPI lights, the plane's airspeed fell to 134 knots, below the **137** set in the MCP but close enough to not trigger any concern – assuming anyone was paying attention.

A few seconds more and the plane was at 331 feet of altitude and 130 knots, descending at a slightly improved 1,000 feet per minute. They were now *under* the glide slope.

Another few ticks of the clock and they were at 219 feet altitude, 122 knots airspeed and *way under* the glide slope. The plane were 0.7 nautical miles – 4,250 feet – from the runway. They were still coming down too fast, 900 feet per minute, and flying much too slowly. Kang-kook continued pulling back on the yoke, raising the nose up first to 7° and then to 9°, which is a lot. With the higher angle-of-attack their speed decreased further.

Kang-kook was probably wondering why his plane was continuing to drop so quickly. The throttle should have kicked in by now to reduce their rate-of-descent, settling the jet on the glide slope. Something was very wrong, and he could not figure out what.

Now Instructor Pilot Jeong-min perked up. 'It's low,' he said, yet again not being clear – did he mean their speed or altitude? It didn't matter – *both* were dangerously low. They were 180 feet above the bay.

'Yeah,' Kang-kook said and kept pulling back on the yoke. The Pilot Flying was out of the loop.

Nine seconds later, 90 feet above the water, Jeong-min had seen enough. Time for action. The Instructor Pilot said, 'Speed' for a second time, and then with his left hand shoved the throttles fully forward to the stops. Takeoff power.

Kang-kook pulled the yoke further into his chest, raising the nose to 12°. But he was not flying a lightweight aerobatics plane, he was flying a passenger jet the size of a small ocean-liner whose engines take time to develop full power and that requires altitude beneath its wings to change its downward momentum.

Kang-kook did not have time or altitude.

The concrete threshold of Runway 28-Left begins at an average of thirteen feet above San Francisco Bay, depending on tides, and right at the seawall. When Jeong-min pushed the throttles forward the plane was less than a quarter of a mile from that seawall and rapidly sinking towards it.

While the engines were spooling up, the airspeed fell to 103 knots and the stick shaker went off, warning of an imminent stall. Jeong-min called out, 'Go around.' The NTSB report doesn't tell us if the Instructor Pilot was calm or shouting.

Then the increased engine power started having an effect. The plane's speed increased to 105 knots and the stick shaker stopped vibrating. The stall had been averted.

But it was too late. The Boeing jet's tail slammed into the seawall, sheering off as the rest of the plane careened down the runway. Miraculously, of the 307 people aboard the plane, only three died. Forty-nine were seriously injured, 138 sustained less serious injuries.

Chapter 23

Not So Complicated

Like with the crash of Colgan 3407, there are two reasons why on a picture-perfect summer day Asiana captain Lee Kang-kook smacked his plane into the seawall of SFO's Runway 28-Left. One is the pilot's immediate act of commission or omission – his action or inaction – that directly led to the crash. The other is the chain of events, the links strung together that explain what brought the plane down.

The immediate reason for the accident was Kang-kook's failure to establish his plane on a stabilized approach. On a day as nice for flying as that Saturday, a stabilized approach was easily doable by a captain of his experience. With it, his autothrottle would not have placed itself in **HOLD** mode and his airspeed would not have decayed to stalling speed.

The chain of reasons Asiana 214 crashed does not have many links. It doesn't extend much earlier than the approach, unless we include a possible Mental Fatigue Hazardous State of Awareness everyone on the flight deck may have been experiencing after the long overnight flight.

Perhaps the most obvious link, as well as the most important, is the Runway 28-Left glide slope beam being out of service. If it had been working, Kang-kook would have presumably flown the approach in the expert way he had flown them hundreds of times before.

The glide slope beam being out of order led directly to Kang-kook's elevated stress level, the second link in the chain. If he had not been 'stressed' about the approach, his decision-making would likely have been better, particularly his autopilot mode changes, and especially the last one – the one he did not even recall making.

Jeong-min's presence in the cockpit is a third link. Without an Instructor Pilot watching him, at some point Kang-kook probably would have bailed out of the approach, taking the safe path and going around to try again, instead of plowing ahead with an unstabilized approach.

But the biggest link, the one dominating the explanation for this crash, is that Kang-kook and Jeong-min had both lost situational awareness.

Kang-kook had lost track of his altitude, airspeed, and – when it counted most – his autothrottle mode. He seemed to have no idea how far he was above the glide slope, nor how to effectively descend to capture it and ride it down to the runway threshold.

Jeong-min had also lost situational awareness, proved conclusively by his erroneous position report in his radio call. His slow reaction to their unsafe descent below the glide slope also points to his lack of SA, and maybe even to a Hazardous States of Awareness slowing his thought processes.

Beyond the chain links, even beyond the loss of situational awareness, there is a better and broader reason for the crash of Asiana 214, one that covers all the links: aviation automation.

While still hours away from landing and miles above the Pacific Ocean, Kang-kook had been nervous about the visual approach to Runway 28-Left. Yet the NTSB found nothing in his past to explain his worry, no failed check rides or instructor warnings of possible deficiencies in his flying ability. Nothing pointed to why he was 'stressed.'

Landings are not easy, but an experienced pilot grows into the skillset needed to bring hundreds of passengers safely down from almost eight miles in the air. A student pilot learning to land a single-engine training plane is taught to manage airspeed, altitude, rate-of-descent, flap settings, landing gear and radios, all while keeping the runway in sight and an eye out for other airplanes. With practice the workload lessens. The pilot still has to accomplish everything on the list, but it somehow feels easier.

As a pilot's training planes get faster and heavier, the list of items to manage grows, the margin for error decreases, and the time available to get things done – and done right – shortens.

Automation helps.

With autopilot handling, for example, the rate-of-descent, the pilot now has extra bandwidth to monitor airspeed and plan for a missed approach. Add a glide slope beam for autopilot to follow, taking the guesswork out of finding the proper angle down to the runway. Then throw in autothrottle to manage airspeed, eliminating that worry. Use enough automation and the workload can be reduced to where the pilot has relatively little to do except watch.

And that is where trouble begins. Complacency sets in, especially since automation fails so rarely. Pilots like Kang-kook know their automation is eminently trustworthy, just as Parasuraman's *constant-reliability* test subjects in his 1993 experiment knew theirs was, as well.

That is true until an emergency or a mistake shocks the pilots out of their complacency. Or until a glide slope is taken out of service.

THE DANGERS OF AUTOMATION IN AIRLINERS

Aviation automation is the true culprit behind the crash of Asiana 214 not only because it put the entire crew into one Hazardous State of Awareness or another. Long before the flight, automation had robbed Kang-kook of some of his basic piloting skills. It had taken him years to develop the ability to juggle multiple flying and monitoring tasks effectively, correctly, and in the right order while flying a plane full of passengers. When he stopped using them all, when automation began making things easy, those skills rusted.

Without even looking at his logbook we know for sure that early in his career Kang-kook could visually judge an approach to a successful landing. All student pilots first fly visual approaches because they do not yet know how to use an ILS glide slope beam.[36] If they can't do it, they don't get their license. Kang-kook had been flying planes since Asiana hired him in 1994. That was 19 years before the crash. He spent his first year attaining his private pilot license and then his instrument rating. He definitely knew how to visually judge an approach back then.

But it is possible that since that first year, Kang-kook's landings were almost always aided by automation like autoflight or glide slope beams. That is plenty of time for his visual-approach skills to dissipate or disappear entirely.

ILS approaches were not created with the expectation that pilots would follow them in good weather and let their visual judgment and skills deteriorate. They were invented so bad weather would not hinder flight operations. It probably never occurred to the inventors that it would also be used by pilots on good days, because…why would they? Pilots enjoy landing planes themselves. It's fun. That's why they became pilots in the first place.

But what was true in the past may no longer be the case. Pilots around the world are being ordered by their company's operating manuals to use every ounce of their plane's automation. Asiana's chief 777 pilot said as much. The NTSB report on the Asiana 214 crash says he 'stated in an interview that the airline recommended using as much automation as possible.' The direct result of that emphasis is an inexplicable failure to land a plane under perfect weather conditions.

It is worthwhile reviewing the approach once more, this time looking at the entire cockpit crew.

In the early stages of the descent, when the plane first rose above the glide slope, Jeong-min might have been giving Kang-kook leeway to work

---

36 I am not using *visual approach* in the legal sense, under Instrument Flight Rules. I mean *visual* as opposed to *on instruments*.

out of the problem on his own. Maybe the Instructor Pilot was taking into consideration that it wasn't a risk to their safety. Being too high at the beginning of an approach was not dangerous because they could always go around.

Or, as I had suggested, perhaps Jeong-min was already complacent and having Task-Unrelated Thoughts and disconnected from the flight.

He confirmed his Hazardous State of Awareness in his transmission to SFO tower, giving their position as two miles further away from Runway 28-Left than they actually were. That is not a trivial mistake. He had lost situational awareness.

Though not an excuse, I can guess why he made the error. The approach chart for Runway 28-Left, which Kang-kook had first used to compute his rate-of-descent showed DUYET's distance as 7.4 miles away, measured by DME – Distance Measuring Equipment. But away *from what*?

DME broadcasts exact distances between a ground-based antenna and an aircraft, displaying them in the cockpit. It is often used on instrument approaches. But for landings, the DME antenna's position relative to the runway threshold is important. For SFO's Runway 28-Left, the antenna was two miles *past* the threshold, meaning Jeong-min needed to subtract two miles when figuring how far they were from touchdown. '7.4 DME' is actually 5.4 miles away from the foot of the runway. If he had forgotten to subtract two – or worse, did not even realize he should have – that is a sign he was out of the loop.

Then 84 seconds later, when they flew right through the glide slope identified by the PAPI, Jeong-min *had to know* they were in serious trouble. Yet again he did nothing for 14 seconds before declaring ambiguously, 'It's low.' If Kang-kook was not going to abort the landing it was Jeong-min's responsibility to do so.

But he didn't.

He was acting as complacently as a proctor monitoring students writing essays on a final exam. His poor overall performance during the approach while on his first flight as an instructor cannot be attributed to lack of experience, that he had not yet learned the right way to correct a student. And it was not cultural either: Jeong-min was Kang-kook's senior in rank, in logbook hours, in age, in every way that mattered.

Their lives were on the line here. Jeong-min should have been more proactive. First he had let Kang-kook take them too high on the approach. Now they were too low. A cardinal rule of approaches is, never ever go low, you can smack into a building or antenna tower. Yet Jeong-min allowed it for 14 seconds, an eternity at that point in the approach. To me, he wasn't merely complacent. He was locked in a classic diminished-alertness Hazardous State of Awareness.

Dong-won, the relief-crew co-pilot sitting in the cockpit jump seat, initially seemed more engaged than Jeong-min. He had noticed the incorrect **172** knot airspeed dialed into the MCP. He spotted it a bit late, but maybe he actually saw it earlier and had given the senior airman, Kang-kook, a chance to find the mistake on his own.

But then he too missed how high they were as they neared DUYET. He had a clear view of the cockpit navigation display with the green line showing the plane would not reach 1,800 feet altitude until far past the waypoint. And he could have done that simple rule-of-thumb math in his head – 300 feet down for every mile. But he said nothing about it to anyone. He was definitely not 'monitoring well.'

His 'sink rate' callout was not helpful either. Kang-kook and Jeong-min knew how quickly they were descending. Dong-won noticed it because it was unusual and even scary, plunging down so fast so near the runway. But it would have been much better and far more helpful had he noticed DUYET, or the autothrottle's shift to **HOLD** mode.

Beyond their obvious Complacency, the NTSB report raised the possibility that the three pilots were suffering from Mental Fatigue, the Hazardous State of Awareness that may be the earliest link in the causal chain. Eleven hours is a long time in the air, even if the primary crew spent only half of that strapped into their cockpit seats. Plus Kang-kook reported only five and a half hours of sleep in the past twenty-four, too few to remain sharp, especially since it had been broken up into a few chunks. Jeong-min had gotten eight hours, but it was not eight straight, meaning he was not at his best either.

Kang-kook's mental condition, his hazardous conscious state during the approach, is easiest to identify. He was Complacent about his height above the glideslope, and Complacent about autoflight, not recognizing the autothrottle mode had changed itself to **HOLD**.

He was also Excessively Absorbed by his failure to stabilize the approach. We got a hint of that tunneling, that too-narrow vigilance, when Dong-won needed three tries to get his attention regarding the 180 knot airspeed limit. We saw it again when Jeong-min said, 'This seems a little high,' and it took Kang-kook a few long seconds to reply.

Then, in the last seconds of the approach, Kang-kook may have become fixated on the PAPI lights, staring at them while pulling the yoke back to raise the nose. The PAPI colors were changing because his jet was not holding the glide slope as expected; it was dropping through it because autothrottle was in **HOLD** mode. He was mesmerized by – Excessively Absorbed by – the PAPI lights that were not going his way.

He never so much as glanced at his airspeed.

Neither did the other crew members. As Asiana 214 intersected with the glide slope, all three pilots in the cockpit were so narrowly focused on the PAPI lights that their drinking straw focus was more like looking through the eye of a needle. None of them wondered what would keep them on the glide slope, because they knew the answer: autothrottle.

But it didn't.

Chapter 24

Bias and Surprise

Automation complacency is involved in one way or another in virtually every crash where automation plays a role. If a pilot types in an incorrect waypoint and fails to double-check it, misses noticing an autothrottle or autopilot mode even when it is clearly marked on a primary flight display, stops looking at airspeed, forgets about a button that had been pressed, or any of a host of other errors, complacency is at least partly to blame.

But with some crashes automation is responsible for more than just complacency. Human factors experts noticed many pilots developing such a high level of trust and confidence in automation that, when given a choice, they pick automation's results over their own intuition and experience, even when automation's results are inferior or wrong.

That is *automation bias*.

A great example appeared in the early days of computerized flight planning. Creating a *flight plan* is how every airplane trip begins. Driving the kids to grandma's house for Sunday dinner does not require much planning. Just make sure the kids bathe, you have your keys, your license, you've bought the pie for dessert and filled the car with gas, then go. But driving from Chicago to Florida takes serious planning: the optimal time to leave to beat local traffic, how many miles to drive before stopping overnight, where to sleep, what time to start out the next day, the best time to arrive, and so on. Thought is involved.

It is the same for planes. For a pilot flying her small plane fifty miles to visit a friend, the *flight plan* – the details of that trip – might be derived by simply drawing a straight line from airport to airport on a Visual-Flight-Rules Aviation Sectional map. The map contains all the communication and navigation radio frequencies she will need. She'll note a few landmarks readily identifiable from the air, like highways and train tracks, lakes and rivers. Then she'll take off and follow that line.

For pilots taking an airliner across a continent or across an ocean, flight planning is considerably more complicated, so computer software has been created to help. While these programs are now available on laptops and tablets for private pilots taking those fifty-mile trips, twenty years ago only commercial aircrews had access to them, running on their airline's mainframe computers.

Studies were done back then to gauge the impact of computerized flight planning on a pilot's route selection. In one surprising result, researchers discovered pilots would accept the computer's routing recommendation even when they knew it was not the best choice for their trip, and even when they had come up with a better route on their own.

That is classic automation bias: going with automation's choice because it comes from a computer. Computers know best.

Much scarier is a pilot's automation bias on a commercial plane with 300 passengers aboard. Pilots can reach a point where they so unequivocally trust their plane's automation that they will assume everything it does is correct. They won't bother checking because they have learned checking is a waste of time – nothing has ever gone wrong, and nothing ever will go wrong. Automation bias.

Look how Kang-kook used automation near the end of the flight to keep his plane flying at 137 knots. He dialed that airspeed into the MCP and assumed everything would take care of itself. He figured the throttles would automatically hold that airspeed no matter what.

When he saw the PAPI indicating his plane had reached the glide slope, he pulled back on the yoke, raising the elevator and raising the nose. He expected the same result he had gotten every other time he had done that: his rate-of-descent would slow while his airspeed held. That is exactly what *would have happened* if autothrottle had been set to the correct mode.

But instead, in **HOLD** mode it was set to do nothing. Kang-kook had assumed autothrottle had his back, that it would do its job. He was *biased* to think that way because it had always done its job in that situation every time.

So at the right moment, Kang-kook pulled back on the yoke, the nose tilted up, and Kang-kook assumed his jet would settle on the glide slope. Instead, the jet descended below it. He pulled back even more, tilting the nose even higher. Nothing in Kang-kook's brain registered surprise, nothing alerted him that something was wrong, not even the extreme nose-high attitude, because autothrottle was presumably working and so the extreme angle was, he figured, normal.

No it wasn't and no it wasn't.

Automation bias faked Kang-kook into thinking automation was giving him the right answer: if the nose was tilted higher than normal, it must be the new normal. But actually, it was not giving him any answer. Sure, his failure to check his airspeed or notice the throttle mode setting was Complacency, which is why we have focused on that until now. It was Excessive Absorption as well. But as you now see, it was even more than that.

If the logic behind **HOLD** mode seems odd to you, the NTSB felt the same say. In its report on the Asiana 214 accident, the bureau pointed to a serious issue raised during certification flight testing of Boeing's 787 in the US and Europe. The airliner, which first flew in 2009, is similar in size to the 777 and has autothrottle mode logic nearly identical to the older jet.

Referring to the 787's **HOLD** mode, the NTSB said both the FAA and the European Union Aviation Safety Agency (the European Union equivalent to the FAA) 'expressed concern about the intuitiveness of this design from a pilot's perspective.'

'Expressed concern' is another way of saying 'we don't like it.'

Despite the aviation agencies' unease, Boeing refused to modify **HOLD** mode's function. Instead, it inserted a note in the 787 Aircraft Flight Manual – something every pilot studies – that read, 'When in HOLD mode, the A/T [autothrottle] will not wake up even during large deviations from target speed.'

Kang-kook might have benefited from that note, but Boeing did not go back and provide the same warning to 777 pilots.[37]

What happens when pilots are doing everything right yet automation takes them down a rabbit hole no one in the cockpit anticipated? What if some important nuance of an airliner's automation was left out of the manual? What if a crew finds themselves facing something they had never practiced in a simulator because no one knew it could happen?

That is called *automation surprise*.

---

37 The NTSB report on Asiana 214 says that 777 pilots could have learned about **HOLD** mode's more dangerous features by a careful reading of the plane's Flight Crew Operating Manual. But it would not have been easy. The NTSB said the information would have to have been 'carefully pieced together by the reader because it is provided in two different notes with dissimilar wording that could be interpreted as describing different issues.' The bottom line is, Boeing could have been far more forthcoming about **HOLD** mode's idiosyncrasies.

Chapter 25

LIFUS

Amsterdam Schiphol Airport is the busiest airport in Europe measured by the number of flights in and out. It is the third busiest measured by passengers. Built during the First World War on what had once been a shallow lake bed, its location is ideal for modern airline route systems connecting European passengers with the Middle East, the Americas, and the rest of the world. One of the air carriers regularly flying in and out is Turkish Airlines, the national flag carrier of Turkey.

On 25 February 2009, just thirteen days after Renslow and Shaw's deadly encounter with automation, a Turkish Airlines 737-800 bound from Istanbul with 128 passengers and four flight attendants had its own run-in with automation that ended badly. But this was not only about automation complacency. This time automation caught the flight crew unawares with a design feature they never knew about that took them completely by surprise. Automation surprise.

Turkish Airlines Flight 1951 took off at 8:23 am local time on a chilly and cloudy Wednesday morning from Istanbul's Ataturk Airport. Schiphol was roughly 1,400 miles to the north-west, typically around a three hour flight. In the cockpit were not only the regular two-man flight crew of pilot and copilot, but also a third pilot, a safety pilot observing from the jump seat.

In command was Captain Hasan Tahsin Arisan, a 54-year-old graduate of the Turkish Air Force Academy. After flying fighters in the air force, he joined Turkish Airlines in 1996. He had 17,000 hours in the air and was now an Instructor Pilot.

Next to him in the right seat was his student, 42-year-old Murat Sezer, another air force veteran and a relatively new hire at the airline. He had 4,146 total hours as a pilot, most of them in his country's air force, the remainder at Turkish Airlines accumulated since joining the air carrier eight months earlier. He was qualified to fly the 737-800 but not ready to work as a copilot – the First Officer – of a passenger flight on his own, so Arisan was doing double-duty as both the flight's captain and Sezer's flight instructor.

As with the relationship between Lee Kang-kook and Lee Jeong-min, describing Arisan as Sezer's instructor was technically accurate but overstating Arisan's role, as well as understating Sezer's abilities. Sezer had already received his *type-rating* for the 737, confirming he was fully and legally qualified to fly it (type-ratings are permission to take command of a specific type of airplane, such as a Learjet, or a Boeing 777). But his final training phase required forty flights with an instructor pilot in the captain's seat as he learned the nuances of 'flying the line' – flying regularly scheduled trips with paying passengers behind him. This is known in commercial aviation as LIFUS, Line Flying Under Supervision. Under their airline's rules the first twenty LIFUS flights require a 'safety pilot' in the cockpit as a second set of eyes monitoring the student, ready to step in if needed. The second twenty LIFUS flights are flown without the safety pilot aboard.

Since Sezer was still working through his first twenty LIFUS flights, sitting in the jump seat behind him and Arisan was 28-year-old safety pilot Olgay Ozgur. Though much younger than his cockpit companions, he had far more airline experience than Sezer, having been fully qualified on 737s for two and a half years, and with 2,100 total hours as a pilot.

With Sezer as the Pilot Flying, Turkish 1951's takeoff from Ataturk Airport was uneventful, as they nearly always are. Two and a half hours later, while 36,000 feet over Germany, Sezer gave his landing briefing to the other two pilots. He told them the weather at Schiphol Airport was overcast and misty, though not actually raining. The base of the clouds was at 1,300 feet with a few clouds at 600 feet and visibility around two miles. Since low clouds obscured their view of the runway they would land using an *instrument approach*, specifically the ILS approach to Runway 18-Right. The temperature in Amsterdam was a chilly 39° Fahrenheit. They would be landing in just over thirty minutes.

Soon they reached Dutch airspace and control of their flight was assumed by Air Traffic Control the Netherlands, which guided them to the final approach course for 18-Right. With around eleven minutes to go they were at 250 knots airspeed and 8,400 feet, making their way down to 7,000 feet.

Suddenly the landing gear warning horn went off in the cockpit, alerting the pilots the gear was not down and locked.

That made no sense.

Normally this warning goes off close to the ground, while the airplane is flying near landing speed and with flaps and slats deployed, obvious signs to the plane's computers that landing is imminent. At that point the landing gear *should* be down and locked, but if it is not, the warning starts blaring.

For Turkish 1951 flying at 8,400 feet, the warning was a mistake.

It is not known whether it occurred to any of the crew that something was amiss because none of the pilots survived the crash. Conversations picked up by the Cockpit Voice Recorder give the only clues to what they knew and did not know as the flight progressed. It happens this crew spoke very little, forcing investigators to guess what some of their more cryptic comments meant. When the loud landing gear horn warning began, none of the pilots said a word.

After ninety seconds the gear-warning horn stopped.

A few seconds later Captain Arisan said, 'Radio altimeter.'

Investigators are not certain, but presumably he was blaming the landing gear warning horn on a faulty radio altimeter, whose altitude output was shown on his primary flight display. If that is what he meant, he would have been right.

The 737-800 carries two radio altimeters, designated left and right. We first came across them when introducing *autoland*. Altitude readings from the left radio altimeter are displayed on the captain's – the left seat's – primary flight display. The right radio altimeter's readings are displayed on the copilot's PFD.

They are highly accurate instruments, but to be safe they have a built-in layer of security. Each radio altimeter self-checks every altitude it reports to the plane's computers, labelling each one *USABLE* or *UNUSABLE*. Most altitudes are *USABLE*, but if a radio altimeter detects an obviously incorrect height, it designates the reading as *UNUSABLE* so the plane's computers won't use it. Then the device is automatically taken off-line as a precaution, and the second radio altimeter is used instead.

At least, that is how the flight crew believed it worked.

Dutch accident investigators dissecting the plane's avionics discovered Arisan's primary flight display was showing a radio altimeter altitude of -8 feet, eight feet below sea level, obviously a ridiculous reading. But mysteriously, the radio altimeter self-checked the altitude as *USABLE*. The 737's computers therefore believed the plane was actually that low, noticed the landing gear wasn't deployed, and in response set off the landing gear warning.

The Dutch investigators learned Boeing had encountered this identical problem with its radio altimeters multiple times over the years, both at Turkish Airlines and at other carriers using the plane. The plane manufacturer could find no discernible pattern to the errors. Bad altitude readings would just happen now and again.

Boeing engineers could not find a reason for the incorrect altitudes, nor why the radio altimeters' software could not consistently recognize an *UNUSABLE* altitude reading for what it was. The company suggested replacing the radio altimeter antennas, installing special mounting gaskets to prevent moisture from seeping into the antennas, and even replacing computers, but nothing made the problem go away.

The issue had never led to a crash, and it was never fixed. Instead it was forgotten.

For Turkish 1951, their faulty left-side radio altimeter had no impact on this early part of the approach to landing since Sezer was doing the flying using only the right-side autopilot along with the right-side radio altimeter, which was working perfectly. The plane has two autopilots, left and right; because of the broken left-side radio altimeter, only the right side autopilot was being used.

But why did Captain Arisan wait one and a half minutes after the warning horn began before commenting on it to his crewmates? Perhaps he was waiting for his student, Sezer, to notice the failure. But the copilot could only see his own radar altimeter and would not have known about Arisan's broken one. It remains a mystery.

A few seconds later the landing gear warning horn went off again, this time for only two seconds. No one said anything.

A full forty seconds after that, while the cockpit was busy but quiet, Arisan spoke again. 'Landing gear,' he said.

'OK,' Sezer replied.

Investigators are not certain why Arisan made this comment now. His earlier 'radio altimeter' callout was somewhat understandable. But his 'landing gear' did not fit anywhere. The plane was too high and too far from the runway to lower the landing gear, so his comment was not suggesting that. He may have been referring to the earlier warning horn erroneously notifying the crew that the landing gear should be down, but the crew had already lived with that annoyance, so he was not telling them anything new. In the end, what Arisan meant by it is entirely unclear.

After a minute, the landing gear warning horn fired yet again for two seconds. No one mentioned this one, either.

While the landing gear warning horn was going off Sezer continued flying the plane, following Air Traffic Control directions now coming from Schiphol Approach Control. Schiphol's last instruction was to descend to 2,000 feet. They were getting close to final approach.

Like Kang-kook on Asiana 214, Sezer was not hand-flying, but rather he was flying by turning and pressing the knobs and buttons of the plane's Mode Control Panel. Though they were aboard a 737-800 rather than a 777-200ER, both jets were built by Boeing and their MCPs were similar. When Schiphol Approach ordered the plane to descend to 2,000 feet, Sezer dialed **2000** in the MCP and the right-side autopilot did its thing.

Chapter 26

I Have

At 10:22 am Amsterdam time Sezer's autopilot levelled them off at 2,000 feet. They had been flying at 250 knots, but now they needed to slow down. Sezer set the airspeed in the MCP to **195**, and autoflight dutifully slowed the plane to 195 knots. Sezer asked Arisan – the Pilot Not Flying – to lower a first degree of flaps. Arisan did as his co-pilot requested.

Less than one minute later, Schiphol Approach Control radioed, 'Turkish 1-9-5-1, turn left heading 2-1-0, cleared approach, 1-8 right.'

'Left 2-1-0, clear I-L-S, Turkish 1-9-5-1,' Arisan replied, confirming they could begin their final approach to Runway 18-Right after making a slight left turn to a heading of 210°.

Sezer used the MCP to slow the plane further, first dialing in **170** knots, then **160**. He also asked Arisan to deploy more flaps and lower the landing gear to help slow them down. Each time Arisan did as instructed.

They were nearly lined up with the runway, though the crew could not see it because they were in the clouds. Sezer pressed the button setting the MCP to **APP** mode – Approach mode. Autoflight would now handle the ILS approach by locking on to both the glide slope and localizer radio beams. The crew had no reason to suspect this instrument approach would be any different than the many they had flown before.

Turkish 1951 was five and a half nautical miles from the threshold of Runway 18-Right, and still at 2,000 feet. Using the 300 feet-to-1-mile landing rule-of-thumb, they were 350 feet above the glide slope. It was time to go lower. To get down, Sezer pressed the **V/S** mode – Vertical Speed mode – button on the MCP, just as Kang-kook had done. Then he twisted the rate-of-descent knob to **-1400**.

The autopilot briskly dropped the plane down at the programmed rate of 1,400 feet per minute. It would stay at that rate until reaching the glide slope, when autoflight would capture it and begin following the radio beam to the runway. Meanwhile Sezer expected autothrottle to hold their airspeed at 160 knots by managing engine thrust.

A Bombardier Q400 in Continental Connection livery, identical to Colgan flight 3407. (Courtesy of Rudi Riet)

A view of damage to the fuselage of Asiana flight 214 after it crashed into the seawall at the foot of San Francisco International Airport's Runway 28-Left. (National Transportation Safety Board)

NTSB Investigators on scene at the crash of Asiana flight 214. (National Transportation Safety Board)

The interior of the shattered fuselage of Asiana flight 214. (National Transportation Safety Board)

NTSB aerospace engineer Greg Smith receives the recorders from Asiana flight 214 in the NTSB's laboratory in Washington. (National Transportation Safety Board)

Turkish Airlines flight 1951, in a field one mile short of Amsterdam Schiphol Airport. (Fred Vloo/RNW)

A distant view of the wreckage of Turkish Airlines flight 1951. (Fred Vloo/RNW)

Turkish Airlines Boeing 737-8F2 TC-JGE at Kiev-Boryspil Airport in 2008. This is the aircraft that crashed as Turkish Airlines flight 1951. (Courtesy of Pawel Kierzkowski)

PK-LQP, the Boeing 737-8 MAX that crashed as Lion Air flight 610, photographed in September 2018, six weeks before the accident. (Courtesy of Ikko Haidar Farozy)

Ethiopian Airlines' ET-AVJ, the aircraft which flew as the ill-fated flight 302, takes off from Ben Gurion International Airport, Israel, in February 2019. (Courtesy of LLBG Spotter)

Lieutenant Colonel Henry Munhoz shows a photograph of a recovered piece of wreckage of the Air France Airbus A330 of flight 447. (Agência Brasil)

Lawrence Sperry and Emil Cachin in flight above Bezons, France on June 18, 1914, while demonstrating Sperry's gyrostabilizer. Sperry is standing in the cockpit. Cachin is standing on the wing, six feet to his right. (Courtesy of *Scientific American*)

The Ford Trimotor, the first plane to fly passengers far enough and fast enough to make the economies of air travel worthwhile for airlines and passengers.

The Boeing 247, the plane that led to the modern concept of air travel. This particular example, shown hanging in the Smithsonian Institute's National Air and Space Museum, is the first production 247-D. (Smithsonian Institute)

The Airbus A300B; the first Airbus aircraft to enter commercial service. This aircraft carries the markings of the first A300B1 prototype, including original F-WUAB registration. (Courtesy of Sebastien Loubet)

The Boeing 737 MAX, the third generation of Boeing's 1960s-era 737 aircraft, after the Classic and NG. (Courtesy of Steve Lynes)

Except autothrottle would be doing nothing at all.

When Sezer set the autopilot mode to **V/S**, he anticipated autothrottle would automatically change to **SPD** mode – speed mode – which would maintain the jet's airspeed. That is how it always behaved.

But not that morning.

Instead, autothrottle automatically went into **RETARD** mode, electronically moving the throttles all the way back to *idle*. **RETARD** mode means 'retard the throttles,' pull them back to idle specifically – and only – because the plane is about to execute the landing flare. Normally a 737 pilot flares around 35 feet above the runway, but if the plane hasn't flared by 27 feet, autothrottle will automatically switch to **RETARD** mode and pull the throttles back on its own. It is just how it's programmed, and pilots should know about it.

But Turkish 1951 was still 2,000 feet in the air. The landing flare was minutes in the future. Something had gone very wrong.

That *something* is autothrottle was connected to the left radio altimeter – the broken one, the one registering eight feet below sea level. If the left radio altimeter's output had been flagged *UNUSABLE*, autothrottle would have automatically switched itself to read the right radio altimeter. But since the left one was flagging its altitudes as *USABLE*, the plane's computers believed the plane really was at -8 feet. So when Sezer punched the **V/S** mode button, autothrottle assumed the plane must now in the landing phase, and **RETARD** was the correct mode for that last stage of the flight.

Both primary flight displays announced that autothrottle was now in **RETARD** mode. It was right smack in front of the pilots, in the upper left of their displays. Yet none of them saw it. More accurately, we don't actually know what the pilots saw, since they did not survive. But if any of them *had* seen the **RETARD** message light, they didn't say anything.

It is possible they saw it but it never registered as meaningful to them. At the time, they were lined up perfectly on the localizer beam, they were above the glide slope beam but approaching it quickly in **V/S** mode, and they were slowing down. Everything was going according to plan.

When Sezer punched the **V/S** button, the crew of Turkish 1951 expected to see the throttles move back to *idle* to help the jet descend.

They saw that.

They thought it was happening because autothrottle had switched to **SPD** mode and was slowing the plane.

No. It was happening because autothrottle had switched to **RETARD** mode and thought they were close to touching down.

111

At 10:24 Schiphol Approach Control radioed Turkish 1951 to contact Schiphol's Control Tower.

Before Arisan could call the tower, the safety pilot, Ozgur, announced to Arisan and Sezer that he noticed a radio altimeter wasn't working. He didn't specify which one.

Arisan acknowledged Ozgur in Turkish with, 'Tamaaaam,' which the Dutch accident report officially translated as, 'Ooookay.' Unofficially it probably means, 'Yeah, we know, thanks for reminding us.' That is all the captain said. Sezer said nothing at all.

They were four nautical miles from the runway.

Arisan radioed Schiphol tower for permission to land, which was immediately granted.

A few seconds before 10:25, they reached 1,300 feet and intercepted the glide slope. Sezer's right-side autopilot now began following the radio beam downward by raising the elevator to lift the nose and reduce their speed and rate-of-descent. Eighteen seconds later they dropped below 1,000 feet. They were out of the clouds and could see the runway straight ahead of them. Everything continued looking just as it should.

Without being asked by Sezer, Captain Arisan set the flaps to 40°, as the jet needed more *lift* as it slowed. Then Sezer changed the MCP speed setting to **144**, the speed he intended to hold most of the way down to the runway.

Twenty seconds after passing 1,000 feet, they reached 750 feet, and the plane slowed to *below* 144 knots. The jet needed engine power to keep its speed up while descending on the glide slope, but autothrottle was not cooperating.

No one noticed.

If Sezer or his cockpit mates had only looked at the airspeed indicator on their primary flight displays they would have realized they were in trouble. But no one looked. And still no one spotted **RETARD** on their primary flight displays either. They were all busy going through the Landing Checklist, believing autothrottle was managing their speed and adding power when needed.

But autothrottle was not adding anything.

They had two miles to go.

Sezer said, 'Gear down, please, three green,' confirming the nose and main landing gear were down and locked in place.

Arisan called out, 'Five hundred,' in Turkish.

Sezer asked Arisan to turn on their landing lights. The plane was ready to land.

I HAVE

They were passing through 500 feet and now flying at 110 knots. The captain was obviously closely watching altitude. But he still was not checking airspeed. Given the proximity of the two instrument readings on his primary flight display – they are only a few *inches* apart – it is impossible to understand how he never glanced from one to the other. But he did not. Neither did Sezer.

Suddenly, at 460 feet and flying at an amazingly slow 107 knots, the stick shaker's rattling stall warning went off, shocking the crew into action. Instantly and correctly responding like the air force pilot he once was, Sezer pushed the yoke forward with his right hand and shoved the throttles up half-way towards full power with his left. He should have pushed the throttles all the way up, but he might have been in mid-push when Arisan suddenly snapped, 'I have,' in English. The captain was taking control of the flight. Sezer removed his hands from his yoke and the throttles.

At the instant the stick shaker started, the plane's nose was tipped up a steep 12° above the horizon. It is staggering to realize none of the pilots picked up on this cue that something was terribly wrong. They had been sitting there like passengers while autopilot had tilted the nose higher and higher. Taking off at this angle is normal. Landing is not. We can only wonder what the three crewmembers were thinking while the nose was rising up.

Now Ozgur said, 'Speed Hocam,' followed quickly by, 'One hundred knots, Hocam,' and then another, 'Speed, Hocam.' *Hocam* very loosely means 'my teacher,' but it was more of a general honorific used liberally among Turkish Airlines' pilots without regard to rank.

Ozgur's callouts were too late. He had utterly failed in his singular responsibility as safety pilot to protect them. He did not survive, so we will never know what he was looking at or what he was thinking about during the approach, but he obviously was not looking at either primary flight display. Investigators ruled out fatigue as a factor – Ozgur and the other crewmembers were well rested. It is likely that he, as well as Arisan and Sezer, were in one of the Hazardous States of Awareness – maybe more than one of them.

Captain Arisan now had the yoke in his left hand, which was good. But he did not have the throttles in his right, which was bad. He had probably seen Sezer push the throttles up and must have thought they had stayed forward, perhaps because he assumed overruling autothrottle would cause it to disengage.

If so, Arisan had assumed wrong.

The **RETARD** mode was designed to *keep* the plane's engines at idle, no matter what a pilot might do. So when Sezer removed his hand in mid-push, **RETARD** mode automatically moved the throttles back to idle, where they remained doing nothing.[38]

You might wonder why Sezer did not see or feel the throttles moving by themselves back to idle. It is because of his training. While learning to fly, when an instructor says 'my plane' or 'I have it,' students are taught to instantly take their hands off the yoke and throttle and hold them in the air – the way Lawrence Sperry did back in his demonstration flights in 1914 – to show the instructor it is truly his airplane, he is in complete control.

Knowing they were facing a dire emergency, when Arisan said, 'I have' – leaving off *it* or *the plane* because of the urgency of the moment – Sezer probably let go of the throttle and yoke as if they were scalding him, raising his hands in the air so Arisan would know for sure he had taken over. Sezer would not have felt the throttles returning to idle and he would not have thought to look down at them either.

By the time Sezer pushed the yoke forward the plane had already stalled and was plunging below the glide slope. Investigators computed the plane stalled at 105 knots.

The only way for Turkish 1951 to recover from the stall was to point the plane downhill and pick up airspeed quickly. Sezer had started that process by shoving the yoke forward. When Arisan took over he continued pushing the nose down until it was 8° below the horizon. They were going downhill, but without engine thrust they were not going downhill fast enough.

Finally, nine seconds after the first stick shaker warning, Arisan spotted the throttles sitting at idle and pushed them all the way forward to the stops. At the same time, with the ground rushing towards them in the cockpit window, he hauled the yoke into his chest. Four seconds later the engines were at full power. The plane's nose was pointed 22° up into the air.

It was too late. Turkish 1951 ran out of altitude, smacking into a soft field belly first, one mile from the threshold of Runway 18-Right. The three pilots died, along with one flight attendant and five passengers. Almost all the remaining passengers and crew were injured, some very seriously.

---

38 When Jeong-min in Asiana 214 pushed the throttles up while in **HOLD** mode, they stayed where he put them. Not so in **RETARD** mode. That's why pilots must know every nuance of every mode.

Chapter 27

Perfect Storm

The immediate cause of Turkish 1951's crash was a straightforward error of omission: the crew's lack of attention to their primary flight displays. If they had been scanning their instruments the way they had been taught since their earliest days as student pilots, they would have noticed their airspeed decaying and either aborted the landing or recognized autothrottle was not working as expected.

That same instrument scan done properly would have brought **RETARD** into clear view. It was not supposed to be there, **SPD** was. Again, had they been properly engaged, even if they missed their falling airspeed they would have seen this.

Automation complacency.

The longer-running cause, the chain links leading to the crash, start with Boeing not working hard enough to fix the intermittent radio altimeter issue of labeling bad altitudes *USABLE*. Turkish Airlines had alerted Boeing to the issue in 2001, eight years earlier. Other airlines had as well.[39]

The plane-builder stopped trying to solve it, figuring 737s have two radio altimeters, so if one goes on the fritz the other can take over and the plane's computers will separate good readings from bad. About the failing radar altimeters, the Dutch accident report writes that Boeing believed 'this did not involve a safety problem.'

It certainly was a safety problem aboard Turkish 1951. Just because it had not yet caused a crash, that was no reason for Boeing to abandon the problem. But the company did.

Turkish Airlines Flight 1951 crashed because of a perfect storm of automation complacency and automation surprise.

---

39 It is possible Boeing knew about this earlier than 2001, but the Dutch accident report only provides information about Turkish Airlines' communications with Boeing regarding the radio altimeter. Also, the radio altimeters were not manufactured by Boeing. The Dutch report says two electronics companies built versions of them, but both had this issue, and Boeing took the lead in trying to resolve it.

THE DANGERS OF AUTOMATION IN AIRLINERS

The flight crew's failure to scan their primary flight display properly was predictable. As Raja Parasuraman's 1993 experiment showed, when pilots trust automation they become complacent and lower their vigilance. Even though something had already gone wrong aboard Turkish 1951 in the early stages of the approach – Arisan's broken radio altimeter – none of the crew expected anything else would go against them. They continued trusting their automation. Complacency.

But what happened next turns out to have been a harbinger of much worse to come for Boeing, on its 737 MAX. Sezer, in the right seat, was the Pilot Flying, controlling the plane with his right-side autopilot. He believed the right-side radio altimeter was working with the plane's computers and autothrottle, while ignoring the broken left one.

Surprise. That was not correct. The 737's autothrottle always used the *left side* radio altimeter, no matter which pilot and autopilot was controlling the plane. It was a holdover from the early days of flying, when captains – always in the left seat – ruled their cockpits like fiefdoms and handled almost all the flying. Even with the right side autopilot in control, the left side radio altimeter was guiding autothrottle, unless it was reporting *UNUSABLE* altitudes.

We know it wasn't. Turkish 1951's left-side radio altimeter was reporting *USABLE* altitudes, so it stayed connected to autothrottle.

According to the Dutch Safety Board report, Boeing never told anyone in any instruction course or operating manual that autothrottle would use the left radio altimeter regardless of which autopilot is in command, switching to the right *only* if the left's output was *UNUSABLE*. Arisan and Sezer therefore did not know autothrottle was reading altitudes from a broken radio altimeter.

A textbook definition of automation surprise is, 'when an automated system behaves in ways that the operators do not expect.'

Pilots cannot expect something that they have never seen before, is not found in manuals and is never taught. The crew of Turkish 1951 were experiencing automation surprise.

Not only was the crew surprised by autothrottle's use of the left-side radio altimeter. **RETARD** mode's behavior surprised them as well. They were not expecting the throttles to return to idle after Sezer had pushed them half-way forward during his attempt to recover from the stick-shaker. When Arisan took over control of the plane he never checked whether the throttles were at full power. Even without looking, if he had just reached down with his right hand, his 17,000 hours in the air would have confirmed for him by feel alone that the throttles were at idle. But he never did, because he did not think he needed to check.

Automation complacency prevented him from checking and discovering the automation surprise.

For all their mistakes, whether from automation complacency or automation surprise, once the stick shaker went off they still had a way out of the hole they had dug for themselves. They did not have to crash.

Boeing test pilots simulating Turkish 1951's last few minutes discovered that when Sezer slammed the throttles forward at the onset of the stick shaker, if he had moved them all the way to full power *and* if the throttles had stayed in that position, the crew could have flown themselves out of trouble. They would have had the power and the altitude to reach the runway, or to go around and try again.

But that is not how the flight of Turkish 1951 ended.

Having now read about two crashes on final approach, one in beautiful weather and the other in clouds and rain, automation's similar role in both is striking. In each of the accidents, autothrottle was not working as expected. But neither aircrew realized it because the very few times anyone in those cockpits checked airspeed, it had been right where they expected to see it. After seeing it there for that very short instant, neither crew bothered checking again.

In their careers, the six pilots in the cockpits of Asiana 214 and Turkish 1951 had nothing but good experiences with automation. Autothrottle always held speed or thrust when asked, and autopilot always held altitude and heading when asked. None of them had reason to suspect airspeed would suddenly go rogue.

But it did.

For an airplane, airspeed is life. Without enough of it, a plane is just pieces of metal, carbon fiber and wiring riveted and glued together. Airplanes can fly without fuel by gliding. They can fly with an electrical failure. They can fly with every instrument in the cockpit going haywire at the same time. They can fly without computers

But they cannot fly without airspeed. Not ever.

What was it about airspeed that caused these eight pilots to lose interest in it?

Nothing. The culprit is not airspeed, it is automation.

In the three accidents we have talked about so far, as the planes' airspeeds neared the stall warning, the airspeed indicators on the primary flight displays changed color and shape. Automation's information function was doing its part, visually warning the pilots things were heading south.

Yet that was not enough to make any of these aircrews pay attention.

THE DANGERS OF AUTOMATION IN AIRLINERS

During the early stages of pilot training, with minimal automation at a student's disposal, everything registers because everything is important – airspeed, altitude, rate-of-climb or rate-of-descent, heading, fuel, batteries, and everything else.

But once commercial aviation careers begin in earnest, pilots discover they can dial their desired airspeed into their autopilot and the plane will reach and then hold that speed. No need to keep checking. Airspeed is always right where it should be.

Until it is not.

Part IV

MAX

Chapter 28

Airliner

At this point you must be wondering whether you will ever step aboard a commercial airliner again. Flying across the country is statistically, and famously, safer than driving to the local mall, but it does not feel that way to most people. Being enclosed in a metal tube miles above the earth hurtling along at nearly the speed of sound frightens many of us.

At least, the argument goes, in a car we have control.

People may think they have control in a car, but that is not accurate. They have no control over the driver in the highway lane next to them who swerves without warning, or the drunk driver motoring straight towards them at night without headlights and on the wrong side of the road. Sitting in the passenger seat or in the rear, they have no control over their driver – what exactly would they do if they saw a car barreling towards them? Reach over and grab the wheel?

A car's safety is illusory.

Skewing the flying public's mindset against air travel is the sheer number of lives lost in a typical commercial plane crash. The deaths of hundreds of people spectacularly and all at once makes newspaper front pages and the lead stories of nightly television newscasts. It is an indisputable tragedy that makes us think twice before getting aboard an airliner again.

But despite the damage to our psyches when hearing and reading of airplane disasters, flying commercially is statistically incredibly safe, far safer than ground transportation. A bit of context helps understand any story, and the story of aviation automation's dangers is no different. Where did the airliner come from? How did we go from the Wright Brothers to the Jumbo Jet and the Concorde? How did we develop the capacity to design and build machines weighing – some of them – over one million pounds, and flying nearly as fast as a bullet?

We will only spend a few pages on the subject, hitting the highlights. Yet when we are done, you'll know why Boeing built the 737 MAX, the involvement of Airbus in the two MAX crashes (they are not, repeat not, at fault, but they played a surprising and unwitting role), and why Boeing chose not to tell its pilots about a flawed computer program on board.

The first really large plane that could comfortably carry more than a pilot and a couple of passengers was built in 1913 by Igor Sikorsky, the same Sikorsky credited with inventing the helicopter two decades later. Born in 1889 in Kiev, Russia then and Ukraine now, by the age of 20 Igor was already designing flying vehicles. At just 23 years old he was hired as the chief aircraft engineer of the Russian Baltic Wagon Factory, a manufacturer of railroad cars and motor vehicles that was expanding into the fledgling aircraft manufacturing industry.

In 1913 he rolled out a four-engine plane that could carry ten people, fly for nearly two hours on one tank of fuel, and reach an altitude of 2,000 feet, spectacular numbers for back then. It might have made more sense to first produce a twin-engine plane, but Sikorsky was simply solving an equation: how many engines would he need to get his big plane design into the air. He initially planned on using two but added two more before its first flight. So the aviation world went directly from one engine to four.

Named the *Russky Vityaz* (the *Russian Knight*), its top speed was only 56 miles an hour, not much faster than an earthbound knight's horse at full gallop. That same year Sikorsky followed the *Vityaz* with the *Ilya Muromets* (named after an actual Russian knight), an improved four-engine plane that could carry sixteen passengers and climb to nearly 10,000 feet. Its top speed was not much better than its predecessor, but it had cabin heat and lighting, plus a bathroom on board, which was convenient. During the First World War, the plane was converted into the world's first four-engine bomber.

German airplane designer Hugo Junkers designed the next major leap in aviation. In 1919 he produced the world's first all-metal transport plane, the Junkers J-13, a low-wing monoplane that looked remarkably like single-engine planes today. His transport could only carry four passengers plus two pilots, but its cruising speed – the speed it can fly its passengers the furthest before refueling – was 99 miles per hour and its range was 870 miles. Passenger airlines sprung up to use it, but planes needed to carry more than four passengers if the airline business was ever going to be profitable.

In 1924 Fokker, another German manufacturer, produced a three-engine plane – a *trimotor* – called the F.VII, capable of carrying twelve passengers. Two years later America's Ford Motor Company unveiled its own trimotor that carried eleven adults but cruised much faster than the Fokker. With the speed and passenger capacity of the trimotors, airline economics now began lining up.

From the mid-1920s on, advances in aircraft design and engine capability led to a multitude of plane manufacturers vying for a piece of the growing industry. Flying boats, which had been around since aviation's early days, entered the long-distance passenger picture, opening up transoceanic travel. The thinking of the time was no different than today: if one of a plane's engines failed in flight, its occupants needed a safe alternative. Today passenger planes flying overwater routes must be capable of reaching land, and a runway, with one engine inoperative. Back then the thinking was, on long overwater flights a plane in difficulty could land on the water, hopefully near a ship where repairs could be made. Or the plane crew itself could carry out repairs with tools carried on board, and then take off again.

Aviation's history is full of breakthrough aircraft, planes that permanently changed the game for everyone. Some are better known than others. The Wright Brothers' Flyer is of course one, as is the *Russky Vityaz*. A tragedy in a Kansas farm field led to a third.

On 31 March 1931 a Transcontinental & Western Air (TWA) Fokker F-10 trimotor (a larger version of the F.VII) crashed in Bazaar, Kansas, killing the eight passengers on board, including Knute Rockne, the famous Notre Dame football player and coach. Rockne's death attracted intense public scrutiny to the accident, which investigators concluded was caused by moisture seeping into the laminated joints of the plane's wooden wings. The joints eventually gave way while encountering the stresses of flight.

The FAA's predecessor, the Aeronautics Branch of the Commerce Department, ordered expensive regular inspections to keep the plane safe while airborne, spurring airplane manufacturers to come up with an all-metal alternative the public and airlines could trust not to fall apart in the sky.

Two years later that airplane arrived.

Combining luxury, speed, and safety, it was an all-metal, twin-engine aircraft that could carry ten passengers at record speeds. Equally importantly, it could maintain its altitude on one engine, a rare feat in those days and a vitally important safety feature when flying over the Rocky Mountains for instance.

That plane was the Boeing 247.

Chapter 29

Boeing

In 1881, one year after Elmer Sperry left Cornell University for entrepreneurial life in Chicago, William Edward Boeing was born in Detroit, Michigan. His father, Wilhelm, had emigrated at the age of 20 from Germany to the US thirteen years earlier, eventually finding work with a lumber trader. After marrying his boss's daughter, Wilhelm proved his own business acumen by earning a fortune through acquiring, trading and developing timber and mineral rights. He died in 1890 at 42, from influenza contracted on a business trip to New York City, but not before securing the financial means to educate his son and namesake, William, in Switzerland and Yale.

A year shy of college graduation, the younger William dropped out to follow in his father's footsteps, hoping to succeed in his own right in the Pacific Northwest. While focused on timber and lumber, in 1915 he had an opportunity to fly as a passenger on a plane, hooking him on aviation. He soon obtained his pilot's license and in 1916 founded the Pacific Aero Products Company to exploit whatever opportunities he might come across.

Later that year, William, with Commander Conrad Westervelt, a close friend who was in the US Navy, designed and produced their first plane, the Boeing Airplane Model 1, a seaplane also known as the B&W. Westervelt left the company the next year when the Navy transferred him east, but William had the financial means to keep the young company going. Orders from the US Navy gave the business legs, and in 1917 he reconstituted Pacific Aero, renaming it Boeing Airplane Company.

The young aviation company grew rapidly, both on its own and through acquisitions, branching out from aircraft manufacturing into hauling airmail and carrying passengers. In 1929 Boeing merged with another rapidly growing aviation company, Pratt & Whitney, to form the most remarkable conglomerate in the history of aviation, calling itself United Aircraft and Transport Company.

United Aircraft and Transport owned an all-star collection of aviation companies. Among them were Boeing Airplane Company (planes), Hamilton Standard (propellers), Pratt and Whitney (engines), Sikorsky Manufacturing (planes including flying boats; Sikorsky came to the US in 1919 and founded his eponymous company in 1923), Stearman Aircraft (planes), and United Air Lines (mail and passenger air service).

In 1934 the US government went after the air mail and air passenger industry following a contract bidding scandal, resulting in the Air Mail Act of 1934. The Act prevented aircraft manufacturers from owning airlines, forcing United Aircraft and Transport to split into three separate businesses: United Aircraft Corporation (now Raytheon Technologies) for all manufacturing east of the Mississippi River, Boeing Airplane Company (the Boeing of today) for all manufacturing west of the Mississippi, and United Air Lines (now United Airlines) for passenger and mail flight operations. William Boeing sold his stock and permanently stepped away.

The year before the company broke up, Boeing was ready with its answer to the Fokker F-10 air disaster. It had already designed and test-flown a fast, all-metal, twin-engine military bomber incorporating the most advanced thinking in aerodynamic design, dubbed the Y1B-9A. So when the flying public desired to move on from wooden aircraft, Boeing had a plane it could quickly convert into carrying passengers. The result was the Boeing 247. The first prototype flew in February 1933, and United Air Lines gave its sister company an order for sixty copies of the new plane.

The 247 *looked* modern. A single wing attached to the bottom of the fuselage – rather than a bi-plane's two wings, or a single wing above the fuselage – is standard today but was rare in those days. The plane's wires, control cables, struts and bracings were out of sight, hidden within its smooth metal skin. The spacious passenger cabin offered six feet of headroom and climate control. Its cruising speed was 150 miles per hour, almost half again as fast as the Ford Trimotor, and 30 miles per hour better than the doomed Fokker F-10. In flight the main wheels tucked into the wings, giving it a sleek appearance and aiding the plane's aerodynamic streamlining. The future had arrived.

To keep up with its competition, TWA tried placing an order for the 247, but Boeing's parent, United Aircraft & Transport, turned the airline away, claiming production capacity issues.

While the gamesmanship made business sense, in the end it enabled the Douglas Aircraft Company, a competing manufacturer, to produce the grandparent of the most successful transport plane of all time. Worried about being left behind, and maybe miffed as well, TWA sought a plane to beat the Boeing 247 in both performance as well as aesthetics. Douglas had it. Founded by Donald Douglas, an MIT-educated aircraft engineer who had grown up the son of a cashier in Brooklyn, it designed and built the 12-passenger DC-1 (DC for Douglas Commercial). Sleek-looking, carrying more passengers than the 247 and faster as well, the DC-1 was the better economic decision for airlines. TWA took delivery of the only DC-1 in existence, and immediately ordered twenty of an upgraded model, the 14-passenger DC-2.

The DC-2 order book quickly trounced the 247's, with its success leading directly to the DC-3. In 1935 American Airlines needed a plane to fit fourteen sleeper births for long overnight flights – the DC-2 was too small for the job. The DC-3 not only filled American Air's requirements, but combined with its military variant the C-47 Dakota it became the most-produced transport airplane in history, with 16,000 copies built, many of them still flying.

During the Second World War, the industrial world's attention turned to warfighting needs, but once peace returned, civilian aviation resumed its advance. The jet engine, developed during the war, offered a quantum improvement in performance over propellers and piston-powered aero engines and airplane designers put it to immediate use.

In 1952 Britain's 36-passenger de Havilland Comet became the first commercial passenger jet to use the new propulsion. But it was flying at speeds and altitudes engineers were not prepared for, and within a year a series of crashes caused by metal fatigue grounded it. A redesigned Comet eventually returned to passenger service in 1958, but in the interim Boeing and Douglas introduced much larger, faster jet planes, the 707 and DC-8, aircraft whose size, speed, range and comfort would change flying forever.

Boeing's 140-passenger 707 flew first in December 1957, and entered service with Pan Am in October 1958. Eleven months later the comparable DC-8 began flying with Delta Air Lines.[40] The 707's earlier launch, coupled with Boeing's willingness to produce multiple 707 variants uniquely suited to each airline customer's specific requirements, put its plane in the sales lead against Douglas, where it stayed.

---

40 The number of passengers each could carry depended on the split between first and economy classes, with an all-economy version of the 707 able to carry as many as 190 passengers. The DC-8 was slightly longer and therefore could fit more passengers.

Chapter 30

Brutal Business

Until now, commercial aviation's growth could be characterized as a long-running competition between manufacturers to produce new and better planes, each model carrying more passengers, at greater speeds, over longer distances, and at higher altitudes. Each advance was important, though not dramatic. But the giant leap by the groundbreaking 707 and DC-8 jetliners, flying hundreds of miles per hour faster and carrying many more passengers than the jet-powered Comet or the biggest piston-powered planes (and fifteen times the number of the innovative 247), left the global airline community salivating for bigger still, faster still.

In 1970 Boeing's 747 Jumbo Jet four-engine behemoth satisfied the desire for size, capable of carrying more than 400 passengers across the Pacific Ocean (until the 550-seat Airbus A380 came along in 2005). In the mid-1970s the British Aircraft Corporation and France's government-owned Sud Aviation joined up to introduce the Concorde to passenger service, a needle-nosed beauty that quenched the thirst for speed by flying at more than twice the speed of sound on trans-Atlantic routes.

But buying jetliners that flew faster or carried more passengers was not the primary goal of airline managements. Then, as now, making money was. Airlines needed profits. Bigger and faster planes reached production only because their manufacturers figured they could be built profitably, while their airlines figured they could be flown profitably. Those profit figures may have been the fantastical concoctions of large egos backed by pliant investors or willing governments. But profit, no matter how miniscule or far into the future, had been a projected possibility at some point for the builder and user of every plane in the sky.[41]

---

41 Arguably everyone knew the Concorde was never going to be profitable under even the most optimistic scenarios, but its profit could be measured in the enhanced national esteems of Britain and France, who undertook its development when the US and Soviet Union abandoned the goal of producing their own supersonic transports.

THE DANGERS OF AUTOMATION IN AIRLINERS

In the early 1960s, aviation bosses encouraged engine manufacturers to come up with more powerful yet more fuel-efficient jet engines. Using them, plane builders produced commercial jetliners needing fewer engines that operated at lower costs. The result was the creation of multiple new airplane markets.

At the small end, below the 707/DC-8 in range and passenger capacity were the three-engine Boeing 727, and the twin-engine Boeing 737 and Douglas DC-9, satisfying the need to move 100 to 150 passengers a couple of thousand miles.

Smaller than those jets and with less range were European products, the British Aircraft Corporation BAC-111 and Sud Aviation Caravelle.

At the opposite end, just below the 747's size and range were the three-engine, 250-passenger McDonnell Douglas DC-10 and Lockheed L-1011.

Designing and building each of these airplanes was a stupendously risky venture. Given the many years it took to go from hypothetical concept to commercial flight, manufacturers were guessing what airlines would need five or more years in the future, while airlines were guessing how their target markets would look by then, and whether their customers would even like the new planes.

The upfront investment to create an entirely new plane was in the hundreds of thousands of dollars during the 1930s, in the millions by the 1950s, and – far outpacing inflation – in the billions of dollars by the late 1960s. With that magnitude of initial capital outlay, manufacturers were practically betting the company each time they rolled a new aircraft design out of their hanger door. Even though new planes were usually introduced with orders already on the books (typically from airlines that proposed the initial requirements for range and passenger-carrying ability), the value of those first orders was never sufficient to repay the company for its massive initial investment. At a minimum it took hundreds of orders and ten or more years of non-stop production for a plane to attain financial success.

Since very few planes ever sold enough copies to return the capital laid out to produce them, their corporate creators often merged to keep going, or else simply disappeared. While Boeing's string of successes gave it staying power, Douglas's jets never sold well enough. In 1967 it merged with McDonnell Aircraft Corporation, best known as a builder of military jets and American space capsules, to form McDonnell Douglas. Though they hoped the combined enterprise could better battle with what had become a Boeing corporate juggernaut, in 1997 Boeing acquired McDonnell Douglas, ending that competition.

European companies were no more immune to the risks than their American counterparts. Junkers was absorbed in 1969 in a merger that many years and corporate iterations later ended up within Airbus. British Aircraft Corporation, builder of the BAC-111 and itself a merger of four British aviation companies, lasted until 1977, when it was nationalized and merged with other UK-based manufacturers to yield British Aerospace, which then became today's BAE Systems. Sud Aviation merged in 1970 with another French government enterprise, Nord Aviation, to become Aérospatiale, and ultimately ended up within Airbus as well. Fokker lasted longer than the others, but went bankrupt in 1996.

The aviation business was brutal. Corporate historians would have difficulty finding another industry with such a long and expensive debris trail. While Boeing emerged as king of the American aviation hill, Airbus eventually landed atop Europe's. Boeing's need to keep pace with – if not stay ahead of – the European company is ultimately why it created the 737 MAX.

So Airbus's past is a history we need to know.

Chapter 31

Airbus

Telling the Airbus story requires returning to the 1960s and '70s, when both the jet airplane market and passenger traffic were growing rapidly, and a host of new aircraft designs were on drawing boards and just entering service.

In the early 1960s, America dominated global commercial aircraft design and production. At the time, the US companies – primarily Boeing, Lockheed and Douglas – owned over eighty percent of the airliner market. Alarmed at the prospect of being frozen out of the industry, in 1965 ministers from France, Britain, and Germany – the home countries of Europe's major plane builders – began meetings to strategize how best to respond.

With all the major US and European manufacturers contemplating new aircraft both large and small, it was hard to see how the Europeans would ever sell enough of these new planes to make money on them. The ministers feared their nations' manufacturers would soon destroy their own finances by competing with both the Americans as well as each other, relegating them – if they survived at all – to collecting crumbs as sub-contractors to the American firms.

On the other hand, if the Europeans agreed to work together, success might come if they identified and filled a hole in consumer demand which the major US manufacturers had missed. It was crucial for the ministers that they make the right market call. They had one shot, a single all-in bet, and if they produced an airplane that did not sell exceedingly well, they were unlikely to find the money or public support to produce a second one.

Two years of intense discussions led to a Memorandum of Understanding, signed in London in September 1967, memorializing the ministers' agreement 'for the purpose of strengthening European co-operation in the field of aviation technology … to take appropriate measures for the joint development and production of an airbus.' *Airbus* was a term then in vogue to describe the next broad category of passenger aircraft which would move large numbers of passengers cheaply – like a public commuter bus. Plus, its general meaning was recognizable in each of their languages.

The plane they agreed to jointly build actually created an entirely new market segment: the medium-range, twin-aisled, twin-engine, widebody jet. Formally announced at the 1969 Paris Airshow, no plane like it existed.

Until the Boeing 747, passenger planes had a single aisle running their length. Putting in two aisles called for a significantly wider fuselage – a wide body – meaning a heavier plane with bigger engines burning more fuel. But airline costs are not measured simply by how much fuel a plane consumes. The best cost measure is on an *available-seat-mile* basis, the cost of fuel – or any other cost – to move a single seat one mile (not the cost to move a *passenger*, but rather the *seat*, empty or occupied; airlines have *per-passenger* measures as well). A twin-aisle airplane would make economic sense if it carried more seats over longer distances for every gallon of fuel than a smaller single-aisle plane.

That was Boeing's motivation for producing the long-range four-engine 747; ditto for the three-engine McDonnell Douglas DC-10 and the Lockheed L-1011. The Europeans decided their Airbus would apply that same principle to a *two-engine* plane.

The world's first twin-engine widebody was dubbed the A300, in recognition of the number of passengers it was designed to carry. Placed in charge of taking the new plane from concept to reality were two Frenchmen and two Germans. From West Germany came a politician, finance minister Franz-Josef Strauss, and Felix Kracht, an aerospace engineer who during the Second World War had designed rocket-powered reconnaissance planes and now headed Deutsche Airbus, a consortium of German aviation businesses.

The French contributed Roger Béteille, a graduate of the elite École Polytechnique and also an aerospace engineer, who gets credit for coming up with the final design of the first Airbus, and Henri Ziegler, at the time in charge of SNIAS, Société Nationale Industrielle Aérospatiale – eventually simply Aérospatiale – the French aircraft manufacturing conglomerate.

Ziegler is a particularly compelling figure, not only because of his French-patriot background, but because, like Elmer Sperry, his son's contributions were as important to the aviation industry as his own. We will come to Ziegler's son, Bernard, shortly.

Born in 1906 in Limoges, Henri received undergraduate degrees in management and aeronautical engineering, and then joined the French Air Force in 1928. Rising to deputy director of the test flight center by 1938, at the outbreak of the Second World War he was sent to the US to procure aircraft and parts for the French battle against Germany.

After returning to France to help organize the Resistance, in 1944 he escaped to England, where he was first put in charge of the Free French Air Force, and then held senior general staff postings. Soon after the war ended he was made chairman of Air France, remaining there until 1954. Following a number of cabinet-level positions, in 1968 he became president of Sud Aviation, which two years later became SNIAS and then Aérospatiale.

By then white-haired, but still trim, he worked tirelessly on Airbus with his son Bernard and his French and German partners to create the plane everyone hoped would save Europe's aviation industry.

As the A300's design progressed, lack of interest by airlines combined with changing passenger market forecasts compelled the Europeans to shrink the plane down to 250 seats, but they kept the model number, now designating it the A300B. To formalize the process, *Airbus Industrie GIE* was officially created on 18 December 1970, owned 50/50 by Germany's Deutsche Airbus and France's Aérospatiale.[42] Strauss, the German politician, was named chairman of the Advisory Board, while Ziegler was made managing director and chief executive officer in addition to his duties as president of Aérospatiale.

The next year Spanish aviation firm Construcciones Aeronáuticas SA (CASA) acquired 4.2% of Airbus, diluting the first two partners. Britain's Hawker-Siddeley and the Netherland's Fokker each contributed manufacturing capability to the effort, though they didn't take ownership stakes. It had become a truly pan-European enterprise.[43]

The Airbus A300B entered regular passenger service with Air France in 1974. But although far more economical to operate than any of its competitors, it initially chalked up only a few sales. Then the industry

---

42 GIE stands for Groupement d'Intérêt Économique, Economic Interest Group, a French form of business joint venture.

43 Though not crucial to our story, here is how Airbus became the company it is today: In 1979 British Aerospace (eventually BAE Systems) bought a 20% stake in Airbus Industrie, diluting the Germans and French though not Spain's CASA. Twenty years later the German and Spanish companies merged, and the following year that business merged with the French to create the European Aerospace Defense and Space Company – EADS – which now owned 80% of Airbus. BAE still owned 20%. In 2006 BAE sold its stake back to EADS, giving EADS 100% of Airbus. Now in total control, EADS changed its name first to Airbus Group NV in 2014 (registered in The Netherlands), then to Airbus Group SE in 2015 (registered in the European Union), and then to simply Airbus SE in 2017.

got walloped by two seismic shocks. First came the 1970s energy crisis precipitated by the 1973 Arab-Israeli Yom Kippur War and the resultant Arab oil embargo. After that came intensifying industry competition culminating in the Airline Deregulation Act of 1978.

The changing environment drove airline managements to seek ways of operating more efficiently, leading Frank Borman, the former American astronaut who was president of Eastern Airlines, to acquire four A300Bs in December 1977 for a six-month tryout. Though terms of the deal were never made public, one industry official said, 'As far as I understand it, the deal to Eastern represents a no cost deal to the airline.' Free was a great price.

Borman was so happy with the plane's operating costs that he ordered 23 of them, with an option for 25 more. That broke the logjam. From then on the A300B sold well, and Airbus was here to stay.

But the European consortium would not survive very long if it only manufactured a single plane model. It needed a family of jets, as Boeing now had with its 707, 727, 737 and 747 airliners. As a first step, Airbus first introduced the A310 in 1982, a slightly smaller but longer-range version of the A300B. Though a valuable fleet addition for some airlines, it wasn't a game-changer. Airbus management needed to concoct a plane that would alter the aviation paradigm, forcing airline customers to see the company as a market leader rather than a follower.

That plane was the A320.

Chapter 32

Fly-by-Wire

By 1980 Airbus zeroed in on the short-haul market as the battlefield where it could win market share from the leading American manufacturers. This segment of the airliner market was growing rapidly as *hub-and-spoke* route systems began dominating commercial aviation. In hub-and-spoke, airlines use smaller planes to feed passengers into major *hub* airports, which then redirect them onto other planes bound for their final destinations. It is what propelled Amsterdam Schiphol to the front ranks of European airports.

Hub-and-spoke gained popularity in the deregulated, high-fuel-price environment of the late 1970s, as airlines sought ways to maximize capacity loads while reducing operating costs. It enabled them to move at least as many passengers as old point-to-point routes while using fewer planes. To be most cost-effective, it was important to have the right aircraft plying these short flights.

In 1981 Airbus finalized plans to produce the ideal plane for a hub-and-spoke world, one it felt would give it the lead in global aircraft sales. Called the Airbus A320 (next in line after the A300B and A310), the jet would be a 150-seat, short-to-midrange plane that would have three key advantages over the competition.

First, it would be more fuel-efficient than every other plane of its size.

The second-largest expense for any airline, behind employee salaries and benefits, is the cost of fuel. Fuel costs are, to a large extent, determined by the engines. Manufacturers building a new plane don't simply pick an engine out of a catalogue and bolt it on. Both its physical size and the amount of thrust it produces must exactly match a plane's requirements. And the way it is attached to the plane must be engineered with incredible precision so the combination of airplane and engine perform the way pilots expect.

The Airbus A320 incorporated the most fuel-efficient jet engines produced in those days. Coupled with aerodynamically-efficient wings, the result was a cost-effective plane where airline bean counters could gauge the fuel cost savings over its biggest competition, the current Boeing 737 models.

Second, its single-aisle cabin would be wider than the competition's.

The obvious advantage of a bigger cabin is more comfort for passengers, who have the final say on the merits of any airliner. The Airbus A320's insides were seven inches wider than Boeing's 737 (as well as the 727), meaning passengers in 6-across seating each had an additional inch of room, plus one extra inch in the aisle. That may not sound like much, but it is the difference between being shoe-horned into a seat, and having a bit of space around the hips and shoulders.

Juergen Weber, former chairman of the board of Lufthansa, expressed a satisfied customer's opinion of the wider cabin: 'The customer acceptance of the fuselage of the A320 is better than the Boeing 737. It is pure physics, very simple, and the customer feels it.'

Third, the A320 would incorporate state-of-the-art *fly-by-wire* technology.

Fly-by-wire was a monumental, generational change in how aircraft are controlled. Airbus would be taking a blind leap across not only a physical barrier, but a psychological one as well. The company could not be certain the commercial aviation world would accept it.

In the early days of flight, control surfaces were connected to the cockpit by cables and rods, a direct *physical* link between the ailerons, elevator and rudder, and the yoke/stick and rudder pedals. If a pilot wanted to climb, pulling back on the yoke yanked a cable connecting the yoke to the elevator, which tilted the elevator up. A pilot could literally feel in his fingers the airflow over the control surface as the plane gained altitude.

Planes eventually grew so large and fast that pilots did not have the strength to overcome the sheer power of the airstream (think of the heavy force of air against your hand out of the car window at 60 miles per hour, then imagine going 500 miles an hour). So hydraulics or electric motors were added at each control surface to boost the pilot's strength into machine strength and make handling easier. But planes still used rods and cables to connect the power-assist devices to the cockpit. When Marvin Renslow incorrectly pulled back on the yoke that tragic evening over Buffalo, he was moving his plane's elevator with cables connected to hydraulics.

Today all small planes, and even some larger ones, are still rigged with cables and rods, because they are simple yet safe. They do not need electricity to operate, so a plane can still maneuver around the sky even if its engine is not running or the electrical system has failed. They are not perfect: cables fray and break, rods crack, and control surfaces jam. But they will continue working even when everything else has gone to zero.

Neither are they foolproof. Planes have crashed when aircraft mechanics rigged cables and rods backwards (a pilot pulling the yoke to start his plane *climbing* would discover his plane *descending* instead, which is, of course, wildly disconcerting). It is rare, but it has happened. It is why every pre-flight checklist includes making sure control surface movements are, in their words, 'free and *correct.*' Not just tilting up or down, but in the intended direction as well.

Physical linkages have drawbacks, especially in larger airplanes, chief among them they are heavy and they need constant maintenance. A 747's rudder cables, running from the cockpit to the tail, are over 200 feet long. Connecting the two pilots up front with all the big-Boeing's control surfaces makes for a dense nest of cables, rods and linkages running under the passengers' feet and through the wings and tail.

Sometime after the Second World War, it dawned on engineers that replacing those bulky cables and rods with electrical wires could save hundreds of pounds. Every movement of the yoke could be translated into electrical signals zipping along wires to move the control surfaces. Better still, if computers were attached to the wires, the computers could help control the plane. A plane rigged with wires instead of cables and rods is a *fly-by-wire* plane.

In a fly-by-wire aircraft, if a pilot wants to climb, he pulls back on the yoke, same as always – nothing about the pilot's job is any different. But this time there is no yanking on cables. Instead, electric signals are sent through wires running the length of the fuselage to the elevator, where either hydraulics or electric servomotors obey the signals and raise the elevator, causing the tail to go down, the nose to go up, and the plane to climb.[44]

Military jet fighters started using fly-by-wire in the 1970s, and the supersonic Concorde used it as well. But no subsonic commercial airliner was designed with it before the Airbus A320. The driving force behind it was Bernard Ziegler, Henri's son.

The younger Ziegler was born in 1933 and, like many other intelligent and influential members of French society, graduated from the prestigious

---

44 A common hurdle with all assisted-power systems that do not directly link the pilot with the control surface is, how to give the pilot a feel for the air on that surface. With nothing but electrical wiring or hydraulics connecting the two, how would a pilot receive the physical feedback to know how hard to pull or push? To solve this, designers added springs and coils, and eventually computers, to reproduce what a pilot would have felt if cables had directly connected the yoke or stick to the control surface.

École Polytechnique. Following in his father's footsteps, he then spent twenty-two years in the French Air Force as a fighter pilot and test pilot, so he was well-equipped to understand the benefits and drawbacks of fly-by-wire. After leaving the military for private industry, he eventually joined his father at Airbus as chief test pilot. He enthusiastically championed fly-by-wire, supported by Roger Béteille, the other senior French representative to Airbus, who by the mid-1970s had become its executive vice president and general manager.

Bernard recognized that along with the positive attribute of fly-by-wire's weight advantage came serious apprehension about an all-electric jet's reliability. Physical linkages somehow *felt* safer than being dependent on the plane's electricity staying on. What if the engine stopped and electric current ceased? What if the plane was struck by lightning? Would it turn into an uncontrollable hunk of metal?

Ziegler knew the answer was an emphatic No. The A320 was designed with multiple motors, computers, and power sources – redundant systems, the engineers call it. And just in case, as a precaution, its horizontal stabilizer trim and rudder were controlled by old-school cables and rods. In a complete loss-of-power emergency, the plane could be turned with the rudder, and could be made to climb and descend with the horizontal stabilizer trim in the tail (by moving the trim wheel) and, of course, by adding or decreasing engine power. Electrical failure was now a non-issue.

Fly-by-wire gave Airbus designers the latitude to create a cockpit that *looked* different than anything pilots had ever seen before. Electrical signals transmitting the pilots' intentions to the control surfaces eliminated the need for a yoke and control column attached to the floor between the pilot's legs. The yoke and column gave a pilot physical leverage while pulling and pushing to move a plane around the sky.

With fly-by-wire the leverage wasn't necessary. Strength had nothing to do with controlling the plane. Instead a video-game-type joystick linked the pilot to the plane's control surfaces. Called a sidestick, one was mounted to the left of the captain and one to the right of the copilot. Now, instead of a yoke in front of them, each had a pull-out table. It made in-flight meals much easier to eat.

Ziegler and Airbus made one last cockpit change. Instead of the cluttered wall of flight and engine instruments in use for decades, on the A320 he installed a *glass cockpit* – six flat-panel computer screens, two side-by-side on the instrument panel in front of each pilot, and two between them, stacked one above the other. With the world rapidly becoming digital, glass cockpits were the logical next step for airplanes, replacing the old analogue dial instruments – *steam gauges*, as the dials are now called. [FIGURE 17]

The A320's flat-panels would contain every bit of information pilots needed to fly their planes, in a neat and orderly layout. The most important 'instrument,' centered in front of each pilot, is the primary flight display, the PFD, incorporating all the information that had, until then, been provided by six separate cockpit gauges. [FIGURE 15] We have already encountered PFDs on Colgan 3407, Asiana 214 and Turkish 1951.[45]

I put *instrument* in quotes because, unlike the old gauges, a PFD isn't a single discrete piece of hardware. It is a *picture* of one. Some planes still use old-style analogue systems to produce data that feed into their primary flight displays. But now most planes with PFDs have an 'air data computer' that takes readings from the plane's sensors – mostly modernized and digitized versions of the sensors and gyroscopes driving steam gauge instruments – and translates them into data shown graphically and numerically on the PFD. A pilot might still refer to the airspeed part of the PFD as an *airspeed indicator*, the altitude part as an *altimeter*, and the compass part as a *heading indicator*. But all three are within the single PFD.[46]

Pilots have very little difficulty switching from steam gauges to PFDs. I learned to fly on steam gauges but have flown planes with PFDs without trouble. Besides, PFDs are easier to read and the digital numbers more precise than the old-style gauges.

Just like when Boeing's 247 arrived fifty years earlier, it was the future again.

---

45 The six main instruments, nicknamed a six-pack, are airspeed indicator, altimeter, attitude indicator (showing the plane's attitude relative to the earth), heading indicator (compass direction), vertical speed indicator (for rate-of-climb and rate-of-descent), and the turn-and-bank indicator (showing the bank-angle of the wings, and whether the plane is slipping or skidding in a turn).

46 For instance, gyroscopes on planes with air data computers and primary flight displays have become digital. Called *ring laser gyroscopes*, they work by sensors detecting rotational movement against lasers, rather than against a mechanical gyroscope.

Chapter 33

Guardrails

Bernard Ziegler's contribution to the A320 went beyond its fly-by-wire skeleton. He also espoused a first-of-its-kind computerized monitoring system that would oversee every control input by the A320's pilots. Using the computers embedded within the plane's wiring, every move pilots made would be supervised by software. If it caught them doing something dangerous to the plane and passengers, like banking too steeply, it would stop them. That is known as *flight envelope protection*. Ziegler called it 'the biggest improvement in flight safety in history.'

That might not be hyperbole.

The *flight envelope* is the complete range of maneuvers a plane can perform without stalling or experiencing structural failure. By definition, a plane leaving its flight envelope is at least partially out of control. Military and aerobatics pilots might depart their flight envelope during certain maneuvers and are trained to get it back. But in commercial aviation, an airliner departing its flight envelope is unquestionably in trouble.

Airbus's flight envelope protection software was, in a sense, a pilot babysitter.

This was revolutionary and breathtaking at the same time: it had never been tried before, and it changed the rules for pilots. No longer would they have the final word aboard their plane – computers would. Bernard Ziegler and Airbus could not be sure how pilots would react to something so radically different from what they were accustomed to – not having absolute authority over their aircraft. What if they needed to fly aggressively to avoid a mid-air collision? Would the system prevent them from turning sharply enough, or climbing or descending steeply enough?

Ziegler again replied in the negative. But he is biased. This time the real answer depended on who you asked, and it remains a point of contention today, more than thirty years after its introduction. Boeing eventually joined Airbus in producing a fly-by-wire commercial jet, the widebody Boeing 777, and installing computer software with flight envelope protection monitoring that airplane's every action in the sky. The US company also switched to all-glass cockpits, though instead of sidesticks it opted to leave in place the yoke-and-column configuration its pilots had been using since the manufacturer's earliest days.

The two companies also chose entirely different philosophies of how to implement flight envelope protection.

Every Airbus plane from the A320 on has what is called *hard protection*. Its onboard computers are guided by narrowly defined *flight control laws* dictating what they will, and will not, allow the pilot to do. For instance, the laws stop the pilot from taking the plane beyond a maximum angle of bank, and past a maximum number of degrees pitch nose-up or nose-down. If the pilot tries banking further or raising the nose higher than allowed, the plane simply won't go there.

Ziegler compared it to guardrails on a sharp road curve. 'The guardrail is not there to help you negotiate the curve,' he said, 'but rather to prevent your car from leaving the road should you not succeed in completing the curve.'

Like a car on that curve, within those guardrails Ziegler and Airbus let pilots ask anything of their plane. But airline pilots worried that in an emergency the guardrails would restrict them from tapping into their own hard-earned skills to save their aircraft. They feared they would be handcuffed.

The fly-by-wire 777 came along nearly a decade after the A320, giving Boeing a chance to research how pilots felt about Ziegler's hard protection. The American company decided to leave the pilot in ultimate command of the plane. Like Airbus, Boeing inserted flight control laws into its plane's computers. But its version, dubbed *soft protection*, transmits every pilot action to the control surfaces no matter how close to, or over, the edge of the envelope that action takes the plane. While these laws don't stop the pilot from doing things, they make dangerous actions more difficult to carry out. The closer the plane comes to departing the flight envelope, the stiffer and heavier the cockpit controls become, letting the pilot know loss of control is very close by. Still, a pilot using enough force will be able to push through the stiffness and exceed the protection limits.

One common feature of both manufacturers' flight control laws is they have built-in exceptions where the laws are relaxed, or even removed entirely. The exceptions are triggered by airborne emergencies or aircraft computer failures, giving pilots extra authority when they need it most.

For both manufacturers, full protection is called 'Normal' – Normal Law at Airbus and Normal Mode at Boeing. A middle ground, with some protections removed, is Alternate Law and Secondary Mode, at Airbus and Boeing respectively. The lowest level, essentially providing no protection at all, is Direct Law and Direct Mode. Behind this, for the most serious emergencies pilots have mechanical backup, cables connecting the cockpit to the rudder and horizontal stabilizer trim that do not require power to operate.

A hypothetical real-world example neatly shows the difference in philosophies. Suppose pilots of an Airbus A320 and a Boeing 777 are each approaching a mountainside. Both need to pull up sharply to clear the ridge and save their plane and passengers.

The A320 pilot would shove the throttles forward to maximum power and simultaneously sharply pull the sidestick all the way back. The plane's software will sense the urgency and pitch the nose up as quickly and steeply as possible within the limits of its Normal Law control laws. While the pilot holds the sidestick full-back (it takes very little physical effort), the software will keep the plane climbing at the steep angle it computed, without allowing the jet to stall as it clears the mountaintop.

The 777 pilot would shove the throttles forward to maximum power and simultaneously yank the yoke back as fast as possible, careful not to pitch the nose up too steeply, causing a stall by exceeding the wings' maximum angle-of-attack. The pilot keeps the yoke back as the plane climbs, fighting against the added stiffness of the envelope protection and ignoring the stick-shaker rattling its stall warning, until the plane clears the ridge.

The two pilots' actions are the same save for one important difference. The Airbus pilot can flick his wrist back instantly and leave it back as long as necessary while the computer manages the plane, prevents it from stalling, and keeps it climbing strongly. The Boeing pilot must use stick-and-rudder skills while pulling on the yoke and fighting against both the stick shaker and the stiffness – both warning of a stall close at hand – as the airplane climbs its way out of trouble.

In 1999 the Airline Pilots Association flew test flights nearly identical to what I just described, in a head-to-head face-off of Airbus versus Boeing flight envelope protection designs. Flown in the skies above each company's Flight Test Facility in Seattle, Washington, and Toulouse, France, the test was overseen and reported on by Captain Ron Rogers, now a retired United Airlines A320 pilot, and flown by a group of commercial airline pilots, two of whom were former US Air Force test pilots. They flew an Airbus

A330 against a Boeing 777, both in Normal Law/Mode (the A330 and A320 use identical flight envelope protection).[47]

Reviewing the test results, Rogers concluded that neither flight envelope protection philosophy was substantially better than the other. Airbus, he wrote, gave 'more consistent performance results ... more quickly,' but in spite of this, his test pilots preferred Boeing's soft protection.

The test pilots were concerned Airbus cockpit crews might encounter, in Rogers' words, 'situations unforeseen by the [Airbus plane] designers where the pilot might need to achieve full aerodynamic capability as opposed to being software/control law limited.' In other words, they felt the A330's response in the test situation was unquestionably effective, but it did not allow the pilot to fly the plane right to the edge of its flight envelope. The fact that the plane didn't need to fly right on the edge was not lost on Rogers, who pointedly added that the pilots were expressing a 'subjective judgment' in their preference for Boeing's soft protection.

And there it was. Rogers' report stated what every pilot was thinking about Airbus flight envelope protection: *in a real emergency, I can get more out of the plane myself than a computer can, and I don't want a computer stopping me from doing my best.*

Maybe.

Most pilots' egos insist they can fly their way out of anything. But Bernard Ziegler knew not everyone can fly like a US Air Force test pilot, so he designed and built a plane that would make any pilot good enough, every single time. He is famously quoted as saying he had built a plane even his concierge could fly. Perhaps he succeeded. The question Ziegler was asking the pilot community was, 'do you trust our software enough to believe its judgement is better-faster-more-consistent than your own?'

That question still gets asked by airline pilots every day.

---

47 These test flights entailed each plane descending towards the ground – known as *controlled flight into terrain* – rather than into a mountainside, as I had described.

Chapter 34

neo

Airbus went ahead with the fly-by-wire A320, figuring that if not pilots, then their managements and the flying public would appreciate state-of-the-art aircraft with built-in 'guardrails' preventing mishandling by their aircrews. Giving the new plane a pre-launch boost was Air France's declaration at the 1981 Paris Air Show of its intention to buy 25 of the not-yet-formally-announced planes, with an option on 25 more. When the formal launch announcement for the A320 came in March 1984, Airbus had orders for more than 80 planes on the books.

The first A320 entered regular service in 1988. Reception was immediately favorable, and its order book raced to catch up to the 737's, which had a twenty-year head start. The plane was a clear and present danger to Boeing.

Boeing's first response came one year before the A320's inaugural flight. The new Boeing 737-400 would use engines comparable to the A320's, resulting in a plane whose operating costs were in line with the new Airbus, even if it didn't have the other advantages. Meanwhile the US company was planning a series of new 737s that would not just match, but soundly beat the A320's costs. It took nearly a decade, but in 1994 Boeing put its answer in the air – the 737 NG, for New Generation.

The NG series still didn't match the wider A320 cabin. Only an entirely new jet could do that, and Boeing was not prepared to take such an expensive step. The 737's fuselage width remained unaltered since the 1950s – the same as the 707 and 727 – enabling Boeing to employ much of the same tooling used to build the earlier jets, lowering production costs. Nor was Boeing going to match the Airbus fly-by-wire setup. It would continue using cables between the cockpit and control surfaces.

But Boeing slung more powerful and fuel-efficient engines under redesigned wings, enabling the 737 NG to fly higher, further and with more passengers than 737 Classic models, the new designation for the older generation of 737s such as the 737-400. It handily beat the A320's cost-per-seat-mile numbers, and orders for the new Boeing poured in.

Now it was the European company's turn to respond. Once again it took a few years, but after a couple of false starts in early 2010 Airbus unofficially revealed its intention to produce an A320 with entirely new engines – the n-e-o, the new engine option – which would have 15% better fuel efficiency than previous models.

The prospect of an A320neo terrified Boeing. The small-plane market's greatest growth was among low-cost airlines. Low cost also meant low margin. For airlines needing to shave expenses anywhere they could, a 15% improvement was a fortune in savings. Boeing had been claiming its 737 NG had an 8% cost advantage over the original A320. Now the A320neo would leapfrog Boeing and seize the advantage.

On 1 December 2010, Airbus officially announced the A320neo. Boeing needed an answer, and fast.

The American company wasn't the risk-taker it had been in the early jet age, when it crushed its competition with its 707 and 747. Plus, it had just rolled out its new 787 Dreamliner, a mid-sized long-range jet that had cost more than $32 billion to develop and bring to market. Most importantly, Boeing felt it did not have time to dither. It would lose hard-won customers if it delayed a response to the neo.

Boeing management decided the best way to combat the Airbus A320neo was to produce a third generation of the 737. It would be ready for customers in six years, which sounds like a long time, but in the plane business it is not. Boeing needed to drive its designers and engineers hard to make that timeline. It would come in a variety of lengths and passenger capacities, from the 737-7 carrying a maximum of 172 passengers, to the 737-10 with a remarkable 230 passenger capacity.

Boeing nicknamed its new plane the MAX.[48]

---

48 MAX is in all upper-case letters, a counterpoint to the Airbus A320neo, which uses all lower-case. I doubt it is a coincidence, though I could find no proof.

Chapter 35

Runaway Trim

The crashes we have discussed so far took place on final approach to landing. That should not be surprising, as the Boeing industry study mentioned earlier showed the combined final approach and landing phases of a flight are responsible for nearly half of all commercial plane accidents with fatalities.

Takeoff is far less dangerous. But when things go wrong during takeoff, they go wrong fast. The margin for error is slim. There is no sky under the plane to give pilots time to consider solutions and try them out.

While autopilot is never used on takeoff, automation is always running in the background of modern aircraft. It can be of immense help to pilots in an emergency, for instance when an engine fails. A pilot facing an engine-out during takeoff must react nearly instantly. It is so dangerous and so important to respond quickly and correctly that it is practiced incessantly in simulators. Today automation handles much of that reaction. The pilot's role is no less important, but with automation helping, it is easier to get it right.

In both crashes of Boeing's new 737 MAX, the trouble started immediately after takeoff. Both crews experienced 'flight control problems,' and their radar returns showed their pilots had enormous difficulty maintaining altitude. Something was wrong with the new jet and investigators needed to find out fast what it was.

The obvious first place to look was at the design of the airplane itself.

As Boeing modified the 737 to compete with the Airbus A320, the company strove to keep all its 737 models handling precisely the same way. That would speed FAA approval of the plane during its certification process and would enable 737 pilots to easily transition from older models of the '60s-era jet. The easy transition would keep costs down for carriers upgrading their fleets from the 737 NG and Classic to the MAX. Crew training costs money.

But over the decades since its first flight, the 737 had grown longer and heavier, with more passengers in the back and more fuel in its tanks. It therefore needed more thrust from its engines. True to its name, the MAX is the biggest of them all, the largest of its models being more than 30% longer, 40% heavier, and with engines producing nearly 50% more thrust than the first 737s.[49] Still, the MAX handles the same as its oldest siblings.

With one exception: when flying at a high angle-of-attack.

We know a high angle-of-attack typically occurs when the plane's nose is pitched up steeply while flying slowly. It is the perfect recipe for stalling because a small additional increase in pitch could take the plane over the tripwire, causing the wings to go from generating *lift* to generating nothing. It can be an extraordinarily dangerous place to fly.

The 737 MAX's powerful new engines were physically so much larger than their predecessors that they were hung from the wings slightly more forward and higher than previously, so that they would not scrape the ground. That caused a problem. Flight tests revealed when the 737 MAX flew at high angles-of-attack, airflow around the bigger engine nacelle in its new position inadvertently created additional *lift*. That unexpected *lift* was like giving an extra push to your child going past you on a playground swing. It caused the plane to pitch nose-up even further, bringing the jet perilously close to a stall, or even stalling it if not caught in time. The inadvertent extra *lift* was also cropping up on steeply-banked turns, causing angle-of-attack problems there as well.

Boeing engineers first tried physical design tweaks to fix the problem, but nothing worked. That left two solutions. First, Boeing could warn pilots that when flying at high angles-of-attack and in steep turns they needed to watch out for the extra *lift* from the engine nacelles, ready to counter it by pushing the nose down. That would be a significant handling change requiring extra pilot training to familiarize air crews with scenarios where the extra *lift* could appear, and then practicing the correct response. Since the extra training would be costly and Boeing might even have to pay for it, this solution was highly unappealing.

Or Boeing could create a computer program to monitor the plane's sensors, on high-alert for this extra-*lift* situation. When encountered, the program would push the nose down automatically, without pilot involvement. It would eliminate the problem entirely.

Boeing chose the latter path.

Company engineers wrote a computer program called MCAS, *Maneuvering Characteristics Augmentation System*, with a single function:

---

49 This comparison is the Boeing 737-10 MAX to the 737-300, the first 'classic' model. The first 737s, the 737-100 and 200, were even smaller, with less powerful engines than the 300.

to push the nose down when the angle-of-attack was too high. MCAS would not be on duty all the time – only if two conditions were met simultaneously: when the autopilot was off, and when flaps were stowed.

With autopilot engaged, its software could manage the potential extra *lift*-kick at high angles-of-attack, so MCAS had no reason to be on alert. But with autopilot off, the pilots would be hand-flying the plane. They needed watching, in case the plane's angle-of-attack steepened too much and the extra *lift* suddenly appeared. That would be MCAS's job.

Stowed flaps meant MCAS would not be working during takeoff or landing. During takeoff and near the end of the landing approach 737s always have their flaps deployed and are often flying at high angles-of-attack. The last thing a pilot would want at those moments is an on-board computer program suddenly shoving the nose down.

To lower the nose, MCAS was programmed to use the horizontal stabilizer trim in the tail. Remember from Chapter 10 that horizontal stabilizer trim lets pilots ease the physical load on their arms when climbing or descending. But its function goes beyond just *holding* the plane's nose up or down. It can *move* the nose in either direction as well. That was the Airbus A320's emergency solution if the plane lost all power: the pilot would use the elevator trim wheel connected by cables to the cockpit to climb and descend.

MCAS would use the same capability.

If the pilot of a 737 MAX suddenly pitched the nose up quickly into a high angle-of-attack to avoid hitting another plane, MCAS would instantly tilt up the front edge of the horizontal stabilizer. Like your palm tilting up when out of the car window, the airflow hitting the underside of the stabilizer would counteract the *extra lift* produced by the engine nacelle during the pilot's sudden pitch-up. It would not push the nose down very much, just enough to prevent the plane from stalling from a too-high angle-of-attack. Meanwhile the pilot would have had no idea MCAS had done anything.

That seemed a perfect solution to the fuselage-nacelle airflow problem. If Boeing wasn't going to take the time, trouble and multi-billion-dollar expense of building a new plane to compete with the A320, computers could keep the 737 flying just like it always had.

But every button, every switch, and every bit of software on every plane in the sky today is supposed to be described in that plane's *Pilot's Operating Handbook* – its manual. MCAS was not. This is the second time Boeing left an important detail out of an airplane manual. The Turkish Airlines crash might have been avoided if the pilots knew their faulty left-side radio altimeter was guiding their autothrottle. Now Boeing was repeating that misstep with MCAS. Measured in lives lost, the consequences this time would be far worse.

Why was MCAS such a huge secret that it was left out of the plane's manual? For the same reason it was created in the first place: Boeing knew including it would have necessitated pilots spending extra simulator time flying scenarios that could trigger MCAS. If 737 MAX pilots discovered software was controlling an important aspect of the new plane's handling characteristics, they would have demanded practice time to feel it out. In at least one case, Boeing supposedly agreed to rebate an airline $1,000,000 for every 737 MAX they had purchased if the jet needed extra simulator time over that airline's other 737s.

After the two 737 MAX crashes, Boeing gave its own reason – really an excuse – for not telling MAX pilots about MCAS. The company said the software was designed so that a failure would look entirely familiar to pilots and they would presumably react properly. Why tell pilots about something new, the logic went, when it would not *look* new? Pilots would not know *why* they were reacting properly, but Boeing figured that was not relevant.

It was made to look like *runaway trim*.

Aboard a commercial airliner with *runaway trim*, the horizontal stabilizer motor suddenly behaves like it has a mind of its own, tilting the stabilizer up or down without the pilot touching the yoke switch or the trim wheel. It is extremely uncommon – American Airlines reported just one occurrence of runaway trim in 750,000 737 flights.

Every pilot flying a plane with electric horizontal stabilizer trim – like the 737 and most other airliners – is trained to recognize and correct runaway trim. The solution is as simple as pulling a circuit breaker or hitting a cut-off switch to the motor driving the stabilizer trim. Without power, the motorized trim can't run amok, though the cockpit's manual trim wheel will still operate normally. But considering how rare it is, for most pilots runaway trim training is stored deep in the recesses of their mind, maybe even too deep to recall when it is really needed.

Twice, four months apart, it was really needed.

Chapter 36

Lion Air

Indonesia is the largest archipelago nation on earth, with 17,000 islands spread over three-quarters of a million square miles. It is also the fourth largest country by population, at 260 million people. Jakarta, its capital, is on the north-west corner of Java, a Mississippi-sized island containing half the country's population. The archipelago is crisscrossed by a dozen domestic airlines and served by a number of international carriers.[50]

Lion Air, formally Lion Mentari Airlines, is today the country's largest air carrier, connecting about thirty-five cities. Founded in 2000, its fleet consists of more than one hundred Boeing 737s flying short-haul routes, and five Airbus A-330s for longer trips.

One of Lion Air's shorter routes is Flight 610, Jakarta to Pangkal Pinang, the provincial capital of an island group 275 miles north of Java. Early on the morning of 29 October 2018, a nearly-new Boeing 737-8 MAX sat at its gate at Jakarta's Soekarno-Hatta International Airport being readied to fly this route with 181 passengers, plus six flight attendants and two pilots.

At 5:18 am the captain and first officer were settling into their cockpit as pre-dawn light filtered through the windows. The temperature that morning was 82 degrees. A light breeze blew from the west. The sun would rise in a few minutes, but a brilliant three-quarter moon still hung in the sky. High clouds and thunderheads threatened over the horizon to the north-west, so before taking off air traffic control assigned the flight a routing that would first take it north-east, away from the rough weather.

In command of Lion 610 was 31-year-old Captain Bhavye Suneja, a 6,000-hour pilot from Delhi, India. Most of his time in the air had been spent in 737 cockpits. Married, and with a cheery smile, he would be the

---

50 Norway rules the largest archipelago in the world, consisting of 240,000 islands. Finland, Canada and Sweden have archipelagos with roughly 30,000 islands. Then comes Indonesia, the largest archipelago *country*. Indonesia's population is behind China and India, with 1.3 billion people each, and the US with 330 million.

Pilot Flying, handling the controls. Sitting next to him was his 41-year-old copilot, First Officer Harvino (like many Indonesians, he had only a single name). Powerfully built and a father of three, he had over 5,000 hours in cockpits, most of them in 737s as well. He would be the Pilot Not Flying.

As passengers were boarding and luggage was being loaded, the pilots ran through their checklists and chatted amiably. This would be their third flight as crewmates. Harvino mentioned to Suneja that he was not even supposed to be there. He had been woken at 4:00 am and asked to fill in on the flight. Not to be outdone, the captain informed Harvino that he was fighting the flu – his coughing can be heard on the CVR about fifteen times before takeoff.

The flight was scheduled to depart at 5:45 am but didn't get to the runway until 6:18. The accident report published by Indonesia's transportation authority (the Komite Nasional Keselamatan Transportasi, KNKT) gives no reason for the delay. After final takeoff checks and receiving takeoff clearance, Suneja smoothly advanced the throttles and the plane accelerated down Soekarno-Hatta's Runway 25-Left.

Thirty-two seconds later he pulled back on the yoke to lift the nose. As the plane tilted skyward his stick shaker began rattling loudly, shocking him into an exclamation. Though he did not know it, the problem was the angle-of-attack sensor on the fuselage's left side, his side, just below his cockpit window. A replacement sensor had been installed incorrectly the day before. It was falsely indicating that the plane was pitching up 21° more than the right sensor, and that it was perilously near stalling. The sensor on Harvino's side was working properly, and his yoke was not shaking.

They pressed on. The flight lifted off.

A few seconds later Harvino, monitoring the cockpit screens, said, 'Indicated airspeed disagree,' followed by, 'positive rate.' He was telling Suneja their plane was climbing normally, but a message on his primary flight display was warning that the airspeeds on each of their PFDs were different.[51]

Like the Airbus A320, the 737 MAX has a state-of-the-art glass cockpit, with four wide flat-panel screens containing PFDs and the plane's navigation and engine instruments. Airspeed showing on the pilots' primary flight displays was faulty. They had no idea how fast they were flying.

'Gear up,' Suneja said in response.

---

51 'Positive rate' is a standard callout by the Pilot Not Flying, letting the Pilot Flying know their plane is climbing and it is okay to raise the landing gear. Shaw called 'positive rate' when Renslow flew the takeoff from Newark in Colgan 3407.

Harvino pushed up the landing gear lever. Concerned about the stick shaker and conflicting air speed data, he asked Suneja if he wanted to return to the airport to land.

Suneja did not respond. He was too busy flying.

As the tires tucked into their wheel wells Jakarta's control tower instructed the flight to switch frequencies and call Terminal East air traffic control center.

Before calling Terminal East, Harvino told Suneja they now had a third problem: a new message on his PFD was alerting him that the altitude readouts on his and Suneja's PFDs disagreed with each other. Suneja's showed the plane was at 340 feet, Harvino's at 570 feet.

Suneja acknowledged Harvino's warning.

The jet continued climbing into the early-morning sky and Harvino moved on, calling the Terminal East controller, who greeted the flight with instructions to climb straight to 27,000 feet.

To better understand what they were dealing with, Harvino asked the controller what altitude Lion 610 showed on the Terminal East controller's radar.

'900 feet,' the controller replied. Right then Harvino's display said 1,040 feet while Suneja's said 790.

Suneja asked Harvino to run the Airspeed Unreliable NNC, the Non-Normal Checklist for when airspeeds seem wrong. Lion Air pilots are expected to have the first four checklist items memorized – if airspeed cannot be trusted, a pilot has to know instantly what to do next without looking it up.

Harvino ducked the order, instead asking Suneja what altitude he wanted. Though they were cleared all the way to 27,000 feet, Suneja could have asked for an intermediate altitude.

Suneja said, 'Yeah ... request uh ... proceed.'

Harvino must have been shocked by the odd response, so he suggested to his captain that they fly alongside the runway, remaining near the airport in case they needed to land quickly. Under the circumstances, the suggestion made sense.

Rejecting his copilot's concern, Suneja told Harvino to request clearance to some 'holding point.' Then he banked left, following the routing they had been assigned before takeoff.

Harvino dutifully keyed the radio and asked the Terminal East controller for directions 'to some holding point for our condition now,' a place in the sky where he and Suneja could fly in circles and figure things out.

Pilots rarely make a request like that, but when they do it is usually to work on some issue they have encountered. It is like pulling your car over on a highway to see if an engine warning light will go out by itself.

Sensing something seriously amiss, the controller asked Harvino for more detail.

'Flight control problem,' Harvino replied tersely.

Counting from the start of the takeoff roll at 6:20 am, Lion 610's flight was two minutes old. In that brief time Suneja and Harvino had been dealing with altitude and airspeed displays that did not agree with each other, and Suneja's yoke continuing vibrating a false stall-warning in his hands. Meanwhile the captain was battling the flu, while Harvino was likely dealing with lack of sleep.

And MCAS was not even involved yet.

All their problems that morning can be traced back to the single malfunctioning angle-of-attack sensor, the one a few feet below Suneja's window. It feeds its data into an Air Data Inertial Reference Unit – ADIRU using its acronym. The ADIRU gathers flight information from the plane's multiple sensors and gyroscopes, then computes and displays everything a pilot needs to know – airspeed, altitude, attitude, location in space, and angle-of-attack – on the pilot's and copilot's primary flight displays. If any sensor is damaged, the ADIRU won't have a complete picture of the plane's position in the sky, and the PFD will either display erroneous or conflicting airspeed and altitude to the pilots, or it might stop providing the data entirely.[52]

Computers and sensors are the building blocks of automation. Automation was failing aboard Lion 610.

As the plane accelerated and climbed past 2,000 feet Harvino asked Suneja for permission to raise the flaps, which had been deployed for takeoff.

Suneja granted permission, and then asked Harvino to take control of the plane. He didn't explain why, but he could not have been having an easy time so far. Maybe he needed a break.

Harvino demurred. 'Standby,' he said, pilot-speak for, *I can't right now.*

Suneja continued hand-flying the jet.

The Terminal East controller asked Lion 610 what altitude they wanted.

---

52 I haven't provided the complete array of the plane's electronics and computers, but enough to understand what the crew faced. To fill in a bit more, between the sensors and cockpit displays are also Flight Control Computers (FCCs), Display Processing Computers (DPCs), and Stall Management and Yaw Damper (SMYD) systems.

'5,000 feet,' Harvino informed the controller after checking with Suneja.

The controller granted Lion 610 permission to climb to their requested altitude and instructed them to continue their left turn, to a heading of north-east ('050 degrees,' he said).

With the flaps now tucked into the wings, something new went wrong. The horizontal stabilizer trim motor in the plane's tail turned on by itself, two times in quick succession, trimming the nose downward. Suneja had not ordered it and was unprepared for it.

The plane dived, losing over 600 feet of altitude before Suneja halted the descent by pulling back on the yoke and engaging the yoke's horizontal stabilizer thumb switch by his left thumb, electrically trimming the plane nose up.[53]

Though he didn't know it, he just had his first encounter with MCAS.

---

53 The yoke's thumb switch was first mentioned in Chapter 10.

Chapter 37

MCAS

MCAS – Boeing's software fix for pushing the nose down when the angle-of-attack was too high – had reared its head for the first time two and a half minutes into the flight. It was connected to the broken angle-of-attack sensor through the ADIRU, the same sensor firing Suneja's stick shaker and messing with the cockpit displays. By design, while the flaps had been extended MCAS had been dormant. But with the flaps tucked into the wings and autopilot not engaged, it had come alive. Operating as programmed, it had reacted to a too-high angle-of-attack, which it thought was caused by the oversized engine nacelles. Even though the real cause was a broken sensor, MCAS did what it was built to do.

In the middle of recovering from the first MCAS-driven dive, Suneja asked Harvino to lower the flaps.

Harvino did, then radioed the Terminal East air controller to ask how fast they were flying.

Lowering the flaps disabled MCAS again, which was obviously a great result. But *why* they were lowered at this point in the flight is baffling. After all, they had just been raised.

Flaps, we know, change the wing's camber, providing more *lift* during slower phases of flight, in particular during takeoff and approach to landing. Once raised after takeoff, pilots have no reason to lower them again until preparing to land. But Lion 610 wasn't going back to Jakarta – Suneja had already rejected that option when Harvino offered it.

Lowering them to arrest the 600-foot decent made no sense either. First, it is not a recommended way to recover from a dive.

Second, it is forbidden to lower flaps when flying faster than a certain maximum airspeed. If deployed above that speed, the airstream might damage them or even tear them off the wings, with catastrophic results for everyone on board. The speed limit for lowering flaps on the 737-8 Max is 250 knots. It is written on the instrument panel just below the landing gear handle, in view of both pilots.[54]

Both Suneja's and Harvino's cockpit displays showed they were flying faster than 300 knots. Even though both airspeeds were probably wrong, the pilots *had* to assume they were flying faster than the flaps could safely be deployed. The controller confirmed that, responding to the query from Harvino that the plane was flying at 322 knots. Now the crew knew: Lion 610 was flying 72 knots faster than the maximum flap speed. It is a testament to the sturdiness of the Boeing jet that it wasn't ripped apart.

Why had Suneja deployed the flaps here? Had he lost situational awareness? His stick shaker was rattling, his flight displays were screwy, and his plane had bizarrely pushed its own nose down into a dive. His mind, clouded as it was with the flu, must have been racing in a thousand directions. Had he momentarily forgotten where they were heading? Was the sudden dive so disorienting him that he was now setting up for landing?

Though he may have lost situational awareness, he had not forgotten about Harvino and the Airspeed Unreliable NNC checklist. 'Memory item, memory item,' he barked at his copilot as the jet resumed climbing.

This time Harvino grabbed the cockpit's *Quick Reference Handbook* – the QRH – a spiralbound book of checklists and other information kept within easy reach in case of emergency, and began thumbing through the pages searching for the checklist.

Suneja spent the next two minutes climbing Lion 610 to 5,000 feet while Harvino continued paging through the QRH.[55] Five minutes after takeoff the jet reached its assigned altitude. Suneja brought the throttles back, reducing engine power while leveling off, standard operating procedure after a climb. The flaps were raised again. The CVR doesn't record Suneja asking to raise them, though perhaps Harvino, as the Pilot Not Flying, moved the handle in response to a hand signal from Suneja.

Then Harvino turned to Suneja. 'There is no air speed unreliable,' he said, admitting defeat in his checklist search. But suddenly he found it.

---

54 Airliners do not have software preventing pilots from lowering flaps and slats at too-high an airspeed. It is up to the pilots to guard against that. Even an Airbus aircraft with hard flight envelope protection has nothing stopping its pilots from deploying flaps while flying too fast.

55 During the climb the plane experienced nine 1-2 second bursts of automatic nose-down trim. The KNKT accident report does not provide the reason for them, but does not attribute them to MCAS. The jet's regular automatic trim system routinely tips the stabilizer up and down minutely during flight. The bursts do not seem to have been consequential since Suneja only responded to them with two slight nose-up trim inputs.

'TEN point ONE,' he shouted, bellowing out the checklist page number – page 10.1 – in his relief.[56] He may even have smiled. He started reading from it.

Harvino:	*Condition. Airspeed or Mach indications are suspected to be unreliable. To identify a reliable airspeed indication, if possible, to continue the flight using the Flight with Unreliable Airspeed. Autopilot if engaged disengages ... Already. Auto throttle if engaged disengage.*
Suneja:	*Disengaged.*
Harvino:	*FD* [flight director] *switches both off.*[57]
Suneja:	*Off.*
Harvino:	*Set the following gear up pitch attitude and thrust. Flaps up four degrees and 75% N1.*
Suneja:	*Yeah.*
Harvino:	*75%*
Suneja:	*Yeah.*
Harvino:	*Already now.*
Suneja:	*Yeah.*

Harvino was questioning whether Suneja had throttled back, which would show up as an *N1* percentage. The N1 percentage measures how close to full power the jet engine is running; lower percentages mean lower throttle settings and lower engine power. Suneja had reduced power when Lion 610 reached 5,000 feet, so Harvino's inquisition was inappropriate, maybe even disrespectful. Besides, the copilot could have glanced at the cockpit's engine instruments to see for himself that Suneja had reduced power.

Tensions may have been building.

Incredibly, during the next four and a half minutes, while Suneja and Harvino continued working through the Airspeed Unreliable non-normal checklist, MCAS was fighting the captain, moving the horizontal stabilizer trim in the tail up *eighteen separate times* to push the nose down. Each time, Suneja pulled his vibrating, rattling yoke towards his chest and used his electric trim thumb switch, holding it in the UP position for as long

---

56 In the KNKT accident report Harvino's words are in capital letters, which I interpret as a shout.

57 The Flight Director shows a pilot, on the PFD's attitude indicator, which direction and by how much to pitch or bank the plane to follow a particular flight path. It can work in conjunction with autopilot, or without it.

as nine seconds at a time to force the nose back up. The battle was being fought to a draw.

During those same four minutes the Terminal East air traffic controller directed Lion 610 to make six separate left or right turns. Turns like those add stress and workload to a pilot in normal flying conditions. For Suneja, the added burden of keeping up with the heading changes must have been grueling.

After the 18th time that MCAS had triggered the nose down – almost ten minutes into the flight – the Terminal East air traffic controller ordered Lion 610 to switch radio frequencies to an Arrival controller who would guide them back to Jakarta. With no solution in sight to their 'flight control problem,' it was the right thing to do.

Harvino contacted the Arrival controller and was told they would be directed back to Runway 25-Left, the one they had departed from ten minutes earlier.

After two more MCAS nose-down assaults and yet another turn ordered by the air traffic controller, Suneja again asked Harvino to take control of the plane. He had truly had enough.

This time Harvino accepted. 'I have control,' he said firmly.

Eleven minutes had passed since takeoff.

Suneja took over Harvino's Pilot Not Flying responsibility of dealing with the Arrival controller, twice mistakenly calling his flight 'Lion 650.' He must have been utterly exhausted.

Harvino suddenly exclaimed, 'Wah, is very…'

He had just had his first run-in with MCAS. But unlike Suneja, while pulling back on the yoke Harvino only briefly used his thumb-switch trim, instead of the aggressive nose-up counterpunching trim Suneja had used. Suneja had not warned his copilot of the need for large doses of nose-up trim, and Harvino was not figuring it out for himself.

MCAS struck again, and then twice more. By the last one – the 26th of the flight – Harvino was pulling back on his yoke with more than 100 pounds of force. And still he was using only a little of the electric up-trim from the thumb switch on the yoke.

Without help from the electric trim, Harvino's strength was not enough. The errant software pointed the jet's nose down one last time, the pilots could not counter it, and seven seconds later Lion 610 plunged into the Java Sea.

MCAS had won. Everyone died.

Chapter 38

Ethiopian Air

On the morning of 10 March 2019, just 132 days after Lion Air 610's final flight, another 737-8 MAX crashed. This one, Ethiopian Airlines flight 302, was bound from Addis Ababa to Nairobi, Kenya, 715 miles to the south.

Addis Ababa, home to nearly three million people and the capital city of Ethiopia, is situated at an elevation of more than 7,500 feet in the Ethiopian highlands. Its Bole International Airport has a main runway aligned roughly east-west, pointing away from the highest mountains close by to the north. On the morning of the flight, a few clouds hung lazily 2,500 feet above the airfield. The temperature was a comfortable 61 degrees.

Commanding the plane was Captain Yared Getachew, a razor-thin 29-year-old who grew up in Nairobi, that morning's destination. Baby-faced and with wide-set eyes, he had enrolled in the Ethiopian Aviation Academy, the airline's training school, straight out of high school, and made captain by age 27. His logbook that day showed 8,122 hours in cockpits, though only 1,400 of them in 737s, 103 of those in the 737-8 MAX. Like Captain Suneja four months earlier, on Ethiopian 302 he was the Pilot Flying.

His copilot was Ahmed Nur Mohammed. A 25-year-old Ethiopian native with a trim mustache, he had spent five years in architecture school before switching careers to aviation. He had only a miniscule 361 hours of total flying time. Of those, 207 were in 737s, one quarter of that figure in the MAX. By any measure the Pilot Not Flying was an inexperienced aviator.

At the foot of Runway 07-Right, engines idling and takeoff checklist completed, Getachew pushed the throttles up on Ethiopian 302. It was 8:38 am. Behind him and his copilot sat 149 passengers and six flight attendants.

The instant the plane's wheels lifted off the runway, the angle-of-attack sensor on Getachew's side went nuts, claiming the plane's nose was pointed up at the nearly vertical angle of 74.5°. Obviously it wasn't, but his stick shaker began rattling in response. Meanwhile the faulty angle-of-attack readings, feeding in from the ADIRU, caused incorrect airspeed and altitude numbers to show on the captain's primary flight display. The copilot's PFD continued to work properly.

Like the Lion 610 pilots, this crew flew on.

With the flight airborne Mohammed checked in with Air Traffic Control Radar, saying they were climbing through 8,400 feet (with the runway approximately 7,600 feet above sea level, at that instant they were 800 feet above the ground).[58]

As the flight passed 1,000 feet above the ground, Getachew engaged the autopilot. A few seconds after it locked on, the plane began, for lack of a better word, twitching. The Ethiopian Ministry of Transport's preliminary accident report describes the movements as, 'small amplitude roll oscillations (± 5° of bank) accompanied by lateral acceleration [the nose moving left and right], rudder oscillations and slight heading changes.' The plane was banking left and right, the nose was yawing, and the rudder was dancing.

Ethiopian accident investigators suspect it was caused by the bad angle-of-attack data coming into the plane's flight computers. The result must have been tremendously uncomfortable for the pilots and passengers, even if the movements were only 'small' and 'slight.'

ATC Radar responded to Mohammed, instructing Ethiopian 302 to climb to 34,000 feet and to turn right, towards RUDOL, a waypoint on their route to Nairobi.

Mohammed repeated the instructions to the controller, confirming he heard and understood.

Getachew asked Mohammed to raise the flaps, which he did. But instead of turning right, Getachew told his copilot to tell ATC they were 'unable and request to maintain runway heading.' Getachew wanted to avoid any turns while he sorted out what was happening to his jet.

Before Mohammed could radio the controller, the autopilot disengaged itself. Investigators think the same computers causing the twitching, disoriented by the faulty angle-of-attack numbers, caused the disconnect as well.

Getachew grabbed his still-shaking yoke, taking back manual control of the plane, then again told his copilot to radio air traffic control for permission to remain on the runway's heading, and to add, 'we are having flight control problems.'

---

58 Standard procedure when establishing initial radio contact with air traffic control is to give both the call sign and altitude.

Mohammed keyed his microphone and told ATC they were not going to make the turn, but Getachew needed to remind him to ask permission to remain on the runway heading. The accident report does not say if Mohammed mentioned their flight control problems.[59]

The controller approved the no-turns heading request.

With the autopilot off and the flaps now up, MCAS was free to operate. The misbehaving angle-of-attack sensor triggered MCAS to push the nose down using horizontal stabilizer trim. Getachew responded with his yoke and the trim thumb switch, pushing the nose back up.

Then MCAS did it again. This time Getachew asked his copilot, Mohammed, to help him trim the nose up. It worked. The crew of Ethiopian 302 won the first MCAS battle.

After a brief discussion the crew decided to flip horizontal stabilizer trim switches to CUTOUT. The cockpit has two of those switches next to the throttles, a primary and backup. When set to CUTOUT, both cut power to the stabilizer trim motor.

To their credit, the crew had quickly surmised the plane was experiencing runaway trim, the exact conclusion Boeing had expected from pilots facing a misbehaving MCAS. Perhaps because Mohammed's pilot license was so new, it was fresh in his mind. Or maybe Getachew and Mohammed were just well-trained. They moved the switches from NORMAL to CUTOUT, killing the power to the horizontal stabilizer trim motor.

Two minutes into the flight, while climbing through 9,500 feet, Getachew instructed Mohammed to again tell air traffic control that they were having a 'flight control problem' and that they would like to stop their climb at 14,000 feet. Mohammed made the radio call, and ATC approved.

Though they had control back, the horizontal stabilizer trim was trying to point their nose down – it was stuck in the position it had been in when they had killed the electricity to it. The pilots were keeping the 737 MAX climbing by muscle power alone, pulling back hard on their yokes, but that was getting more difficult by the second. They needed to adjust the horizontal stabilizer trim to nose-up, but without electrical power they had to turn the cockpit trim wheel manually.

They could not.

Even working together they weren't strong enough. The problem was the throttles: Getachew had never moved them back from takeoff power.

---

59 The Republic of Ethiopia Ministry of Transport interim accident report did not provide a full transcript of the cockpit voice recorder, only a summary, with occasional words in quotes.

With all that thrust, airspeed had built up to faster than 350 knots Indicated Airspeed – the speed on Mohammed's correct primary flight display. They were flying faster than the plane's maximum operating speed limit.[60]

An *overspeed* warning clacker in the cockpit had been banging away, alerting the crew that they risked structural damage from the excessive airspeed, but they may not have heard it over the din of Getachew's stick shaker. Their plane was moving through the air so fast that the sheer force of airflow over the tail was holding the horizontal stabilizer tightly in place, like a vice grip clamped on it. Human power alone could not budge it.

Feeling he had run out of options, Getachew decided to return to Addis Ababa's Bole International. He told his copilot to ask for a routing back to the airport.

They had been in the air for three and a half minutes.

'Radar Ethiopian three zero two,' Mohammed said, 'Request vector to return to home.'

ATC responded with orders to make a right U-turn. Getachew banked the plane towards the airport.

While turning, the pilots still needed to gain control over their jet. Only electrical power could overcome the intense air pressure and move the horizontal stabilizer to nose-up. Or slowing down by pulling the throttles back. Correctly recognizing the problem, but ignoring the throttle option, the crew reset the horizontal stabilizer trim switches to NORMAL so the electrical trim thumb switch on their yokes would work again.

They used it to raise the nose once more.

---

60   If 350 knots sounds too slow to be a 'maximum operating speed,' that is because airspeed in aviation is not a simple number. Two versions meaningful to us are True Airspeed and Indicated Airspeed. True Airspeed is how fast a plane is *actually* moving through the air. It is used in navigation. Indicated Airspeed is the speed a pilot sees on the cockpit airspeed indicator. Airspeed indicators compute speed by measuring air pressure against the fuselage: the faster a plane flies, the more air pressure against it. Pressure is affected by altitude and air temperature. To navigate, pilots must make adjustments for those two parameters to convert Indicated Airspeed to True Airspeed. The higher the plane, and the greater the temperature change, the larger the adjustment needed. Computers can make that adjustment instantly, but humans cannot. So only Indicated Airspeed is used for speed limits – like maximum flap deployment speed. Same for air traffic control – they only use Indicated Airspeed when instructing pilots to fly at a particular airspeed. To help this make sense, here is an example: a plane flying just above the ocean on a spring day whose airspeed indicator shows 350 knots Indicated Airspeed is actually flying at 350 knots True Airspeed as well: there are no adjustments needed. But in the freezing cold thin air of 35,000 feet, a plane flying at 350 knots Indicated Airspeed is actually moving at around 650 knots True Airspeed (about 750 statute miles per hour).

But turning on power to the stabilizer trim motor gave MCAS access to the horizontal stabilizer as well. This time the software moved the tail's horizontal stabilizer nearly all the way to full-nose-down trim. Mohammed and Getachew pulled on their yokes as hard as they could, but they were powerless fighting against the combined force of the nose-down trim and the extreme airspeed. With engines screaming, Ethiopian 302 plummeted into a farm field at 500 knots, digging a crater 33 feet deep.

Everyone died instantly.

Chapter 39

ADIRU

The crashes of Lion 610 and Ethiopian 302 took 346 lives. But the flights could have ended differently. Suneja had been successfully battling MCAS, and the crew of Ethiopian 302 had made the right runaway trim call. Perhaps if Harvino hadn't taken over, or Suneja had lowered the flaps sooner in preparation for landing back at Jakarta, or maybe if Getachew had reduced engine power and slowed their plane, the airliners could have returned safely and landed. But in the end MCAS persevered and prevailed.

No one, not even Boeing, disputes the immediate cause of both 737 MAX crashes. The planes were brought down by their MCAS software, hidden from view and operating against the pilots, a clandestine saboteur intent on murder through acts of both commission and omission. The commission was MCAS moving the horizontal stabilizer up to make the nose go down. On each flight it commandeered the plane and tirelessly fought its pilots' attempts to counter it, eventually accomplishing its task by marshalling more force than the pilots could overcome.

The omission was Boeing's telling no one MCAS was lurking. The two crash accident reports, by independent national transportation agencies on different continents, are emphatic about this.

MCAS was the worst kind of automation. It was invisible and unstoppable. As you read already, Boeing put MCAS in the 737 MAX so the plane would handle the same as the 737 NG and 737 Classic. In theory, if MCAS failed, it would fail by doing *something familiar* to a pilot trained on one of those earlier jet models. But that *something familiar* – runaway trim – was actually not familiar at all. It was so rare as to have been forgotten by one of the crews. And even when the second crew correctly guessed runaway trim, they were not able to rescue themselves.

MCAS has rightly been demonized. But looking at the longer-running chain of links reveals it alone did not bring down those planes. It had an accomplice: the 737 MAX's interconnected sensors and computers – its automation.

THE DANGERS OF AUTOMATION IN AIRLINERS

On takeoff and during the initial climb-out phase of the flight, long before MCAS played its part, onboard automation was failing. The pilots had no idea how high they were, nor how fast they were flying. Half their software thought they were pointed steeply into the sky and close to stalling. Half did not. Aviation automation generally, by itself, shares responsibility for these crashes.

Thirty seconds or so after pushing the throttles up to takeoff power, Captain Suneja in Lion 610 and Captain Getachew in Ethiopian 302 both pulled back on their yokes to lift their nosewheels off the runway. Almost immediately, at the most perilous phase of takeoff – too fast to lower the nose back to the runway but with barely enough airspeed to gain altitude – they faced loudly rattling stick shakers.

Automation surprise.

It is a wonder they didn't each have heart attacks.

Then their copilots reported conflicting airspeed and altitude data. This is *long before* they raised the flaps and felt the first nose-down horizontal stabilizer trim moves from MCAS.

On each plane, the stick shaker and the bogus airspeed and altitude data came from the same source: the left-side angle-of-attack sensor. It had failed – automation had failed – and the pilots didn't know it.

The 737 MAX has two angle-of-attack sensors, one on either side of the fuselage just back of the nose. They continuously measure and feed their plane's angle-of-attack to an Air Data Inertial Reference Unit, the ADIRU.

The 737 MAX has two ADIRUs, Left and Right. They are true computers, taking in and interpreting air data from multiple sensors scattered around the outside the plane and turning that raw data into useful and accurate numbers for the pilots' primary flight displays. The Left ADIRU is wired to left-side sensors and sends its output to the captain's PFD, the Right ADIRU is connected to right-side sensors and the copilot's PFD.

There is a maxim in the computer business: *garbage in, garbage out*. If any of the sensors produced bad data, the ADIRU would not be able to compute accurate flight information for the pilots. But how would the ADIRU know it was getting bad data?

It would depend on how many sensors it was reading.

Aircraft using two sensors of any kind often have software programmed to alert the pilots if the sensors don't agree with each other. The ADIRU would not know *which* sensor was wrong, so the pilots would be warned only that one of their displays was incorrect. With three sensors, in a disagreement, software literally takes a vote, and the two equal data points overrule the third.

ADIRU

Why didn't Boeing install three angle-of-attack sensors on each side of the 737 MAX instead of only one?

According to one Boeing engineer, three was a 'nonstarter.' The earlier Classic and NG models had only one per side. Three would have been the sort of major change requiring FAA approval and adding time to the MAX's certification process, time Boeing didn't have in its race against the A320neo. So the MAX remained with one angle-of-attack sensor per side, each working independently of the other.

MCAS could have been designed to read the angle-of-attack data from the plane's two sensors, comparing them for accuracy. But it wasn't. Instead, it read data from only one angle-of-attack sensor linked to one ADIRU. MCAS began every day using the Left ADIRU and left-side angle-of-attack sensor. Then each flight that day it automatically switched sides – to the right, then to the left, then back to the right, and so on. The next morning it would start off using the left-side sensor again.

How could the Left ADIRU on Lion 610 and Ethiopian 302 have known their planes were not actually 21° or 74.5° nose-up? How could the stick shaker or MCAS have known?

They could not.

Pausing for a moment to revisit Turkish 1951, one of Boeing's attempted solutions to their radio altimeters' mislabeling of ridiculous altitudes as *USABLE* had been to compare the output of both radio altimeters aboard the 737. If they didn't match, the autopilot would disengage. Yet Dutch investigators learned that even that 'comparator' (as that report called it) function made an occasional mistake. In their report's words, 'system control software with a "comparator" has not fully eliminated the undesired "retard flare" mode.'

Automation isn't perfect. Even well designed solutions to problems leave problems.

On the doomed Lion Air and Ethiopian Air flights, the Left ADIRU was given bad data from its left-side angle-of-attack sensor. Using that data, it computed wrong airspeeds and wrong altitudes. Since it was the first flight of the day, the Left ADIRU also passed the bad sensor's false angle-of-attack data on to MCAS, though MCAS initially did nothing with it, because the flaps were deployed for takeoff.

To complete the picture, Boeing had offered buyers of the 737 MAX an option, a warning message alerting pilots if the plane's two opposite-side angle-of-attack sensors disagreed with each other. It was like a better car stereo, or leather seats rather than cloth. Neither Lion Air nor Ethiopian Air purchased it.

Even if the airlines had bought the optional warning, I think the outcomes would have been the same. It would have merely made the pilots more confident their stick shakers were wrong and their planes were not near stalling. It is probable they would still have met, wrestled, and lost to MCAS, because they did not know it existed.

Investigators do not know why Ethiopian 302's left-side angle-of-attack sensor failed so dramatically. Conjecture is a bird struck it on takeoff, skewing its output to nose-up-74.5°. We may never learn the real reason, given the total destruction of the aircraft when it smashed back to earth.

We do know about Lion 610's left-side angle-of-attack sensor. The Indonesian accident report not only points an accusing finger at it, it goes further, arguing the plane should never have flown that fateful flight. It should have been grounded.

In four flights over the previous three days, the left-side primary flight display had been showing false airspeed and altitude readings – and on three of the flights the captain's PFD showed nothing at all. The jet was practically brand new, having been certified as safe and ready to fly just ten weeks earlier. Something was broken.

During a scheduled stop on 28 October in Denpasar Airport (Ngurah Rai International Airport) in Bali, Lion Air maintenance crews figured out that the source of the primary flight display failures was a faulty left-side angle-of-attack sensor, and they replaced it. But they installed it incorrectly, with a 21° error built in (bias, rather than error, is the word in the accident report). It apparently was never properly checked for accuracy.

That makes for yet another link in the accident chain, another accomplice, human this time: Denpasar's maintenance staff.

The night before Lion 610 took off, the plane flew a 600 miles trip from Denpasar to Jakarta as Lion Air flight 1043. As soon as its wheels left the ground, the captain, who was the Pilot Flying, experienced everything Suneja, Harvino, Getachew and Mohammed faced, instruments that didn't agree with each other and a pilot's-side stick shaker rattling away incessantly.

The captain, who before takeoff had briefed his crew about the replaced left-side angle-of-attack sensor, quickly crosschecked his and his copilot's primary flight displays and realized only *his* were wrong. He ordered the copilot to take over.

The copilot grabbed the yoke, but after raising the flaps he struggled to keep the jet climbing, telling his captain the plane was 'too heavy to hold back.' In response, the captain suggested using the electric trim button on the yoke to make it easier to keep the nose pointed up – advice Suneja had failed to give Harvino.

The captain also noticed that each time his copilot stopped using the electric trim, the plane responded by trimming itself nose-down. The captain surmised they were dealing with runaway trim, as Boeing had hoped – though we know they were fighting MCAS.

As the Pilot Not Flying, the captain executed the memorized steps from the Runaway Stabilizer Non-Normal Checklist, which included flipping the cockpit's horizontal stabilizer trim switches to CUTOUT, disabling the electric trim motors, exactly as the Ethiopian 302 crew had done. He also ran through two other checklists in the *Quick Reference Handbook*, the Airspeed Unreliable NNC and Altitude Disagree NNC.

MCAS was now incapacitated, but while executing the other NNCs the captain reset the trim switches to NORMAL. MCAS forced the horizontal stabilizer trim to nose-down again, so he set the switches back again to CUTOUT, where they stayed for the remainder of the flight. Since Lion 1043 was flying far slower than Ethiopian 302, the pilots could manually move the trim wheel as they needed, and they landed safely in Jakarta.[61]

After landing, the captain provided Lion Air's maintenance staff with an incomplete picture of the problems aboard the plane. He mentioned the disagreeing airspeed and altitude displays, but omitted the stick shaker and the horizontal stabilizer trim switches (investigators did not report why the captain skipped them). Also after landing, one of the pilots set the switches back to NORMAL, so maintenance noticed nothing unusual in the cockpit. The Lion 610 accident report suggests that had the post-flight report of Lion 1043's captain been more complete, the aircraft would never have been assigned to a flight on 29 October. Instead, it would have been repaired properly.

That is speculation, of course.

---

61 News reports credited a deadheading pilot in the cockpit jump seat with figuring out the problem was runaway trim, and the solution of flipping the stabilizer trim switches to Cutout. The KNKT Final Accident Report on Lion Air 610 credits only the captain with the decision that saved that plane.

Chapter 40

Captain's Choice

Aboard Lion 610 and Ethiopian 302, MCAS played its deadly roles *only* because of identical fateful decisions by the captains of those planes. Before MCAS struck, aircraft automation set off stick shakers and flashed primary flight display warnings that airspeed and altitude displays disagreed. Yet Captains Suneja and Getachew chose to continue on to their destinations instead of returning to their departure airports.

What prompted these pilots to keep going?

Being a commercial airline pilot is one of the most stressful jobs imaginable. Normally, if something goes seriously wrong at work you can stop what you are doing, take a bathroom break, call your spouse, your mom or your friend, reboot your computer, or get a cup of coffee. That is, you can hit pause and assess. If something goes wrong at 37,000 feet, you either fix it or you die. And not only do you die, but so do a few hundred people who had put their lives in your hands. That calls for a certain personality type.

Pilots are self-sufficient. They have to be, since they have no one to ask anything – except in dual-pilot cockpits, maybe the pilot sitting next to them. They are goal oriented. Their thinking is based in reality. In situations where outcomes can be driven by either emotion or analytics, they are likely to fall on the analytical side. That's good, because when the plane is coming apart around them, we want them acting coolly and rationally. They trust checklists to guide them. And they don't care for shades of gray, opting instead for yes/no, on/off, go/no-go.

The sum total of a pilot's personality means, among other things, when the goal is keeping to a flight schedule, no pilot will deviate without an excellent reason. That is especially true when blowing the schedule also means disappointing one's mother.

News reports said Captain Getachew on Ethiopian 302 had spoken to his mother before boarding his flight to his hometown of Nairobi. Maybe he had planned to visit her? Maybe she was cooking him dinner that evening?

He would not have wanted to disappoint her by not showing up, so perhaps he felt personally committed to keeping the flight heading towards Nairobi.

Captain Suneja aboard Lion 610 seemed to have no pressing personal motive for getting to Pangkal Pinang. But he was fighting the flu and must have been looking forward to the end of his shift. He likely felt an obligation to his employer to get his passengers where they had paid to go. Since it was a short flight Suneja probably also had another trip that afternoon. We don't know what his second flight was that 29 October, but we know Lion 610 was already running late, having pushed back from the gate thirty minutes behind schedule. If Suneja was returning to Jakarta, perhaps his desire to get home and back into bed compelled him to keep his plane airborne and heading towards its destination.

There is a better reason neither plane captain considered turning around earlier than they did. It is found in the Lion 610 accident report's review of the flight the night before, Lion 1043. That captain acted exactly as a pilot should, making go/no-go decisions based on hard information and checklists. The report says, 'The Captain's decision to continue to the destination [Jakarta] was based on the fact that a requirement to "land-at-the-nearest-suitable-airport" in the three Non-Normal Checklists was absent.'

In other words, because the checklists did not specifically order them to LAND NOW, the captain saw no reason to return to earth. I have talked to commercial airline pilots and ex-fighter pilots, and they all agree. Whoever designed the Not-Normal Checklists gave them serious thought. If runaway trim or airspeeds and altitudes disagreeing were valid reasons to land immediately, that instruction would have been there in bold letters. But it wasn't.

Besides, pilots facing 'flight control problems' want altitude under their wings to assess what they are dealing with. They have no way of knowing whether, for instance, when they lower the landing gear, something catastrophic will happen. Suneja had faced MCAS for the first time just ninety seconds after the wheels left the ground, when it pitched the plane nose-down and the jet plunged 600 feet. How could he be sure it would not strike again during the final approach back to Jakarta, when they were at, say, 400 feet?

Suneja's decision to climb to 5,000 feet and fly to some 'holding point' was procedurally the right call. Same with Getachew's request to maintain the runway's heading and not turn. I am sure Suneja would have rather turned around and gotten back into bed. Harvino too. But as pilots following procedures and the dictates of their personalities, gaining altitude and performing the aviation equivalent of pulling over to the shoulder was the thing to do.

I have an additional reason why these three plane captains did not initially turn back. They knew they were commanding the most advanced Boeing in the sky. Loaded with computers, it had never failed them. Nor had the older models. They knew as long as they faithfully worked through their checklists their planes would get them where they were heading and bring them back to earth safely.

They had been trusting automation since early in their flying careers. They were *biased* to believe their trust was well placed. Why stop now?

Part V

Landing Gear Up

Chapter 41

Airmanship

There is no debating that automation diminishes pilots' skills. It steals the time they might be spending refining and honing their collection of learned responses used to control an airplane that, in total, amount to airmanship. It also hides weaker pilots behind its computerized excellence, so that airline managements don't realize some of their cockpit crews are not up to the job until an in-flight crisis overwhelms them.

Experts constantly debate whether certain groups of pilots are worse flyers than others. Accident statistics show that in years past pilots in some corners of the globe, flying for new and underfinanced airlines, had accident rates that left much to be desired. But today their safety records are as strong as any airline's anywhere. And no matter where they are domiciled or who flies for them, airlines at the bottom of the list can blame their issues on weak management and the age of their equipment rather than the quality of their cockpit crews.

These days a pilot's age might be a legitimate determinant of airmanship. Younger pilots lean on automation far more, and from an earlier phase in their training, than older pilots who learned in the steam-gauge days. A Boeing employee confirmed it in an internal memo regarding an aspect of the 737 MAX. He was referring to an unnamed specific emergency-response skill that pilots once had. The memo said, in part, 'I fear that skill is not very intuitive any more with the younger pilots and those who have become too reliant on automation.'

It is unfair to make the sweeping statement that commercial pilots of one country or one age group are less skilled than those of another. Some might truly be better (because in everything in life, some person is always best), but none stand out as markedly weaker. Look, for example, at two pilots you already met.

Colgan 3407's Marvin Renslow, US born, US trained, flying for a US airline, pulled back on his yoke at exactly the time he should have pushed forward, and he and his passengers died. On the other hand, Lion Air 610's Bhavye Suneja, born in India, only partially trained in the US, and flying for an Indonesian airline with an atrocious safety record, battled notorious-killer MCAS software to a draw. His ability to counter every MCAS move while keeping his jet roughly on course *and* working through a checklist with his copilot took remarkable airmanship. His piloting skills were impressive and he deserves recognition.

Suneja is an exception to the poor airmanship of the pilots whose performances we have reviewed so far in these pages. But the pilots and flights discussed here represent an immeasurably tiny slice of the flights flown in the past years. Around the globe more than 100,000 commercial flights take off and land every day. Add in military and general aviation flights, and perhaps 50 million times a year planes take to the skies. In this book I cover nine incidents and seven crashes over ten years. In that time there have been perhaps half-a-billion takeoffs and landings.

Since we are on the topic of statistics, the US-based National Safety Council reported that as of 2017 the chance of dying in a plane crash were one in 188,000 during one's entire lifetime. That distinction, *during one's entire lifetime*, is important. It is not the odds of dying every time you step aboard a plane. It is the cumulative likelihood over all your years. When you board a flight for vacation, or business, or to introduce the kids to your great-aunt, on any single trip your chances of being on the wrong aircraft on the wrong day are considerably lower than 1:188,000 (that is, less likely to happen). By comparison, the chances of dying from a lightning strike during your lifetime is similar, at 1:218,000. On the much-riskier side, the odds of being killed in a car crash during your life is 1:103. That is around *two thousand times* more likely than a plane crash or lightning strike.

Commercial pilots are, almost without exception, extremely capable. There are 290,000 of them in the world today. They each spent years working their way up the ladder, adding to their knowledge and experience at each rung, to the point where they mastered the skills to handle a 500,000 lb airplane with 200 or so passengers aboard. Less talented pilots have long before been weeded out through the regular process of testing and retesting.

Still, weak pilots get through, as we know.

We are going to look at a few more flights as we enhance our understanding of automation's downside. One shows a cockpit crew at their worst, as poorly performing as any imaginable, a few more exceptions to the quality and professionalism of the airline pilot community. The others show air crews at their absolute best, defeating automation's attempts to control and confuse, and keeping themselves and their passengers safe.

Air France Flight 447 is up first. It interests me because it shows how one minute of minor automation failure, a single brief hiccup, can turn a flight into a fiasco. Much has been written about the flight and the crew's behavior. The pilots aren't alive to share their thoughts and defend their actions, so all we now have is the flight data recorder and their conversations recorded by the CVR, which are cryptic, confusing and sometimes nonsensical. I am going to stick to facts and add a few opinions of my own when called for. But first let's document what took place aboard the big Airbus.

Chapter 42

South Atlantic

At 7:10 pm local time on 31 May 2009, Air France 447, a pristine four-year-old Airbus A330, started its engines and pushed away from its gate at Rio de Janeiro's Galeão International Airport for an eleven-hour flight to Charles de Gaulle Airport just outside Paris (all times for Air France 447 are given in Rio time). It was a typically warm, even sticky evening in Rio, 75° with partly cloudy skies. The plane was fully loaded with 216 passengers, 9 flight attendants and 3 pilots.

In the cockpit leading the flight crew was 58-year-old captain Marc Dubois, with 10,988 total flight hours in his logbook, more than half as captain on one jet aircraft or another. Among the passengers sitting behind the handsome pilot was his girlfriend, a former Air France flight attendant who had spent the weekend layover with him in Rio.

The right seat was occupied by copilot Pierre-Cedric Bonin, a 32-year-old with a shade under 3,000 hours in cockpits. He was not only a fully qualified air transport pilot, he was a glider pilot as well. That should have signified a deep understanding of aerodynamics and weather: as gliders have no engine, their pilots use rising air currents and deft airspeed management to stay aloft. But on this flight at least, that would not be true. On the outbound flight he had brought his wife along to Rio for the weekend, leaving their two young boys back home in France. She too was sitting in the passenger cabin.

Because of the flight's duration, an extra pilot was along to spell crewmembers. Designated a copilot, 37-year-old David Robert had roughly 6,500 hours in his logbook, including an impressive 39 previous round trips to South America in the past seven years. He had recently switched to airline management and was now working in Air France's operations center. He had taken this round-trip to Rio to maintain his flying 'currency,' logging the required minimum hours per month to retain his legal status as a commercial airline pilot. He would be – or should have been – an important addition to the cockpit while Captain Dubois was on his rest break.

At 7:29 the Airbus took off, its powerful engines effortlessly pushing it up to its initial cruising altitude of 35,000 feet. Young Bonin was on the controls as the Pilot Flying, and Captain Dubois, the Pilot Not Flying, had the radios. The airliner settled into the quiet hum of high-altitude cruise, the passengers eating dinner while the pilots watched the miles click past on their navigation displays. Autoflight was in control.

To reach Paris the flight was first pointed at the equator, where it would traverse the Intertropical Convergence Zone, or ITCZ. The winds and meteorology of the globe's northern and southern hemispheres collide in the ITCZ, often producing dangerous weather. Crossing it was not especially risky, but it was not to be taken lightly either.

At 9:30 pm the crew received a weather update message through ACARS from Air France's Operations Control Center detailing a 245-mile-wide 'convective zone' ahead of them containing 'a cluster of powerful cumulonimbi' clouds.[62] Taken from the crash report written by France's Bureau d'Enquêtes et d'Analyses, or BEA, the quoted words are weather lingo for probable severe thunderstorms and turbulence. Twenty-four minutes later the plane crossed over the sand dunes of the Brazilian coastal town of Natal and flew into the South Atlantic.

At 10:36, while still a few miles south of the ITCZ, a wary Bonin leaned forward in his seat to adjust his navigation display, reducing the radar's range but increasing the clarity to take a closer look at the sky in front of them. He announced to Dubois, 'So we've got a thing straight ahead.' He did not elaborate, but the *thing* was almost certainly thunderstorms showing up on their weather radar less than 150 nautical miles in front of them.

'Yes, I saw that,' Dubois answered.

The BEA investigation's review of cockpit conversation revealed Bonin's nervousness about the storms in their path. A few times he mentioned wanting to climb higher than 35,000 feet to hopefully fly above them. But the plane was, for now, too heavy to gain more altitude. To get higher and stay there it needed to be lighter by burning off fuel, and a lower outside air temperature.

---

62 ACARS, Aircraft Communications Addressing and Reporting System, is the same air-to-ground communication system Renslow and Shaw on Colgan 3407 used to learn their landing speed.

Besides, Captain Dubois had no interest in climbing just then. He would deal with rough weather when they were closer to it, and not before.

For the next six minutes cockpit conversation varied between piloting and personal. Then Bonin, still nervously eying the weather radar, declared, 'There's another one behind.'[63] The plane was sixty or so miles closer to the ITCZ than when he first commented on *the thing straight ahead*, and now he had seen a new storm cell coming into range of the plane's radar.

Dubois said nothing.

Perhaps aware that he was showing his edginess, Bonin tried lightening the mood. At the instant they crossed from the southern to the northern hemisphere, he said to his cockpit mate, 'Right, we're going past the equator. Did you feel the bump?'

'Eh?' The question probably surprised Dubois.

'You felt the bump?'

'[*Expletive*] … no.'[64]

So much for that.

Turbulence began rocking the plane. It was not especially strong, but after a minute Bonin dimmed the cockpit lights and turned on the landing lights. He wanted to see if they were flying in clouds or clear air. To me, he was no different than a nervous passenger sliding open the window shade to peek at the sky as turbulence jostles the plane. At night in a cockpit, flying through clouds, or with cloud cover overhead and below obscuring the stars and ground, it is pitch black outside the windows. But Bonin wanted the primal comfort of knowing what they were flying through.

With landing lights on, he said, 'It looks like we're entering the cloud cover.' It was clouds associated with the storms they were rapidly approaching. Then he added, 'It would've been good to climb now, eh.'

'Yeah if it's turbulence,' Dubois conceded without committing to anything.

The BEA report on the accident summed up Bonin's state of mind at this point. Referencing the copilot's numerous comments about weather and climbing higher, most of which I have left out, the report says,

---

63 The English translation of Air France 447's cockpit conversation is often imperfect or stilted, but is presumably as close to what was said in French as the BEA could manage.

64 The French version of the transcribed CVR recording uses a (!) symbol when one of the pilots cursed. The English version omits nearly all of the curses, instead using the symbol (…), which it defines as 'a word or group of words with no bearing on the flight.' I will indicate where, in the French version, curses were uttered.

'[Bonin's] various interventions ... showed a real preoccupation, beyond the simple awareness of operational risk. Some anxiety was noticeable in his insistence.'

To accuse a pilot of showing 'some anxiety' is a serious charge. And 'beyond the simple awareness of operational risk' might even imply he was afraid. Pilots are human beings, as we know too well, but they are the last people aboard a plane expected to feel fear. Concerned on occasion, perhaps. But fearful, or even just anxious, is highly unusual for a commercial pilot in high altitude cruise.

Chapter 43

Airspeed

The turbulent air around Air France 447 smoothed out after a few minutes. At 10:56 pm, Captain Dubois called Robert, the back-up copilot, out from the crew rest compartment, so he could begin his own break. The senior pilot had picked a curious time to leave, knowing they were closing in on storms, even saying to Bonin, 'it's going to be turbulent when I go for my rest.' Still, he rose from his seat.

Just past 11:00, after Bonin briefed Robert on flight details, the handover was complete. Robert now sat in Dubois' left seat while Bonin continued handling the jet. To stand up on his way out of the cockpit, Dubois had moved his seat as far back and leftward as it could go. Curiously, when Robert sat down he belted himself in with the lap belt, but did not move the seat forward again to comfortably reach the controls. He must have assumed it was not worth the hassle, as he would only be observing and not actually flying the plane.

At 11:06, light turbulence began nudging them again. Bonin called the flight attendants to suggest they and the passengers take their seats, saying the bumps would get rougher in two minutes. He signed off by telling them he would call again when they were out of it.

Two minutes later Robert said, 'Don't you maybe want to go to the left a bit?'

Though they had not specifically discussed it, knowing they were too heavy to gain much altitude Robert had concluded that deviating to the left or right of the storms ahead would be a better course of action than trying to outclimb them. Now was the time to make that move.

'Excuse me?' Bonin's mind had been elsewhere, probably glued to the radar image. He was apparently already in a Hazardous State of Awareness, excessively absorbed with the storm cells ahead of them. Maybe he was even thinking about his kids at home, on the other side of those storms. Task-Unrelated Thoughts.

'You can possibly go a bit to the left,' Robert repeated, this time sounding more like an order than a suggestion. Bonin altered the autopilot's heading by 12° and the big Airbus banked left to its new course. Dubois was apparently correct in showing a lack of concern. Robert had things under control.

Thirty seconds later, Bonin became alarmed by a smell permeating the cockpit. Thinking at first it was coming from the cockpit's air vents, he and Robert had this exchange:

Bonin: *[Expletive] You did something to the air conditioning. But not to the air conditioning.*

Robert: *I didn't touch it.*

Bonin: *What's the smell now?*

Robert: *It's…it's ozone*

Bonin: *It's ozone. That's it. We're alright.*

Robert: *It's ozone. That's why.*

Bonin: *You can feel already that it's a lot hotter.*

Robert: *That's what's hot and ozony.*

They sounded like third-graders. You can practically hear Bonin giggle in relief. Yes, of course they were alright. Passengers in jets slicing through storm clouds at high altitudes can often smell ozone gas produced naturally in the electrically charged air. His anxiety must have been obvious to Robert.

The copilot's relief was soon tested again.

A disconcertingly loud noise entered the cockpit, that of their A330 slashing through billions of tiny rain droplets and ice crystals floating in the thin air of 35,000 feet.[65] Bonin reacted appropriately: he slowed the plane slightly and turned on engine de-icing equipment to prevent the big turbines from jamming with ice crystals and shutting down. While a perfect reaction, he and Robert did not realize the same ice crystals were rapidly clogging the plane's three pitot tubes on the fuselage beneath them.

Attached to points around the underside of the A330's nose, pitot tubes are sensors used to compute airspeed. They are shaped like pistols, their barrels pointed forward and their handles welded to the plane's fuselage. They measure the pressure created by air rushing into the barrel opening. The ADIRU takes that pressure data and translates it into the plane's speed.

---

65 This interpretation of the sound comes from the BEA report, which recorded the impressions of experienced pilots who had listened to the CVR in an attempt to identify the sound. Some said ice crystals, some said rain. Likely it was both.

Airbus A330 pitot tubes had a history of trouble, and Air France had been replacing them with an improved model. Flight 447's pitot tubes had not yet been replaced, and at 11:10 pm two of the three stopped reporting. Though it is impossible to know for sure, investigators suspect ice crystals were clogging the tubes faster than they could be melted.

It was a temporary problem.

The plane's autopilot, which was being fed the pitot tubes' airspeed information via the ADIRU, suddenly faced a complete lack of data it could trust. Programmed to immediately disengage under those circumstances, it did so with three loud beeps and a warning message on the cockpit primary flight displays. The autothrottles disengaged as well.

The time was 11:10 and 5 seconds.

Reacting to the disconnects, Bonin seized the moment. 'I have the controls,' he said assuredly, and wrapped his right hand around his sidestick controller.

'Alright,' Robert replied.

The pilots knew nothing about the clogged pitot tubes. All they knew was autopilot and autothrottle had disengaged. Their PFDs claimed the jet had almost instantly decelerated from 275 knots Indicated Airspeed to an absurdly slow 60 knots, and had simultaneously dropped 330 feet, to 34,670 feet of altitude. It is physically impossible for that to have actually happened so quickly, but that is what their instruments showed.

The plane rolled slightly to the right, probably due to turbulence. Bonin pulled his sidestick back around halfway, raising the nose. He simultaneously flicked his sidestick almost all the way left. Then he shoved the stick nearly full right, then back to the center.

Now an electronically-synthesized voice came through the cockpit's speakers. '*Stall Stall*,' it blared, alerting the crew to an impending stall. Sixty knots – the speed the plane's computers were seeing, even though it was not the plane's true airspeed – was way too slow for the big jet's wings to keep flying and it triggered the warning.

Robert said, 'What is that?'

BEA investigators do not know what he was referring to. It may have been to the stall warning, though he should have recognized that sound. He had heard it in training dozens of times. Or he could have been referring to Bonin's overcontrolling the sidestick, moving it much more than necessary. At high altitude and airspeed, plane controls are far more sensitive than when lower and slower, yet Bonin had slammed his sidestick from side to side like he was playing a video game.

Instead of replying, Bonin once again whipped his sidestick left and right while continuing to pull it back. Then he said, 'We haven't got a good, we haven't got a good display of speed.' He jerked the sidestick left and right a third time.

His series of rapid side-to-side sidestick movements are peculiar, with no obvious reason for them. But one pilot I interviewed suggested a purpose: perhaps Bonin was trying to confirm his PFD's attitude indicator was working properly.

The attitude indicator is the large central image on the PFD, showing a plane's physical positioning in the sky: nose up or down, wings level or banked. [FIGURE 15] Pilots rely on it to maintain spatial orientation. Bonin's left-right flicks were rocking the wings, which would show up on his PFD if it was functioning. Perhaps with those flicks Bonin was confirming for himself that at least one part of his PFD was accurately portraying the state of his aircraft.

Meanwhile Robert, as the Pilot Not Flying, was tasked with quickly figuring out what was going wrong. He had probably already started studying the cockpit display screens looking for clues.

On Airbus aircraft the two flat-panel screens between the pilots are called ECAM displays, for Electronic Centralized Aircraft Monitoring. The screens show engine and fuel information, as well as messages arriving from the plane's computers and sensors warning of problems and system failures. If there are steps for the crew to take to solve the problem, ECAM will display those steps in checklist form.

Summarizing aloud the few cryptic messages on the ECAM, Robert said, 'We've lost the speeds.' Then, still reading ECAM, he added that they were in 'alternate law protections.'

Robert was alerting Bonin that his airspeed indicator was probably wrong. He was also warning his crewmate that the Airbus flight envelope protection, which Bernard Ziegler had made sure was in every Airbus since the A320 to keep them safe from pilot error, was no longer working. If either pilot wanted to stall the plane, they could. Computers would not stop them.

While it was Robert's job to alert Bonin about the loss of flight envelope protections, Bonin could have learned the same thing himself by looking at his primary flight display. Certain PFD colors and details change when the plane switches to alternate law and the protections disappear. But it is possible that in his agitated state of mind the changes would not have registered. It was good Robert mentioned it.

The copilot moved the sidestick left and right a fourth time, still keeping the plane's nose pointed up. Both pilots' altitude readings on their primary flight displays showed they had not only returned to 35,000 feet, they were 500 feet above it and blasting higher, with the nose pointed up like on takeoff.

Only fifteen seconds had passed since Bonin said 'I have the controls.' In that brief period both pilots' airspeed indicators had stopped working, the stall warning had gone off, they had lost flight envelope protections, they were climbing fast, and the jet was rocking from turbulence and Bonin's sidestick movements.

Though they knew their airspeed displays were incorrect, they had no idea why. Nor did they know why the autopilot had disengaged. None of the ECAM messages told Robert what was really happening, that the jet's pitot tubes had become temporarily clogged with frozen ice crystals and were not operating properly.

That is all. Nothing else was wrong with their aircraft.

Still reading the unhelpful ECAM screen, Robert now asserted some control on the flight deck, saying, 'Wait, we're losing ...' He didn't finish the thought. Instead he switched to a new one. 'Wing anti-ice,' he said as he turned on de-icing equipment covering the jet's wings. This means he had either correctly associated the jet's troubles with the ice crystals they had just flown through, or he had feared things might get worse in the frigid high-altitude clouds. He still did not realize the specific issue was the clogged pitot tubes, but he was on the right track.

Robert now said to Bonin, 'Watch your speed watch your speed.'

Once again, investigators do not know what he was referring to. It could have been to their vertical speed, how fast they were climbing, which appeared accurately on their primary flight displays. Or it could mean he had realized – as any pilot would – that the climb was slowing the jet's airspeed (they were going 'uphill'), which according to their PFDs was already at a ridiculously low 60 knots.

Bonin picked up on Robert's intent. He pushed his sidestick forward and said, 'Okay, okay, I'm going back down.'

As the nose began coming down, Robert encouraged him: 'Stabilize.'

Bonin agreed. 'Yeah,' he said, but then he didn't. He raised the nose again.

Now Robert demanded, 'Go back down. According to that, we're going up. According to all three you're going up so go back down.' Robert was probably pointing to the two primary flight displays and a small backup display on the instrument panel, all showing the jet climbing.

Bonin acquiesced, saying, 'Okay,' and pushed his sidestick forward again. He held it forward for roughly five seconds, not enough time to lower the nose below the horizon. Their Airbus continued climbing.

Right here Robert's airspeed indicator resumed working properly. It had been thirty one seconds since the autopilot disengaged.

Robert again: 'You're at … go back down.'

Bonin. 'It's going. We're going back down.' But that wasn't true. They were still nose-up and climbing.

'Gently,' Robert urged. Then he pressed a button summoning Dubois from the crew rest compartment behind the cockpit.

The senior copilot was annoyed. He had practically ordered Bonin to descend, the junior pilot confirmed he would descend, but then inexplicably pulled back again on his sidestick, raising the nose. They were now at 37,500 feet, nearly the very limit of the jet's capabilities, and still going higher. Robert needed the captain.[66]

It was now only forty-five seconds since Bonin had taken over, yet the two pilots were feeling they had almost completely lost control of the jet. But they hadn't. Every control input Bonin was making was being translated into nose and wing movements. The only instrument trouble they were facing was a temporary loss of airspeed indications, and only on Bonin's side; Robert's was working again. They still had plenty of time and altitude to save themselves and their passengers.

Seconds after Robert summoned Dubois, the *Stall Stall* warning voice got going again, and would continue chanting on and off for most of the remainder of the flight.

Then the plane began buffeting. Robert let out a curse.

Buffeting is as real as stall warnings get. I mentioned near the beginning of our story that stalls announce themselves through cockpit horn warnings, stick shakers, and buffets. We had encountered the first two in other crashes, and the horn here, in the guise of a synthetic voice (the A330 does not have a stick shaker). Now Robert and Bonin were experiencing stall buffet. The big Airbus was flying so slowly relative to its angle-of-attack that the wings were clawing for air and the entire plane was shaking with the effort.

---

66 The plane would not have been able to remain at this altitude because, as mentioned, it was too heavy and the air temperature was too warm. But it could *ballistically* climb for short periods like a cannon shot, using its momentum to climb before settling down to a lower altitude again.

In response, Bonin mashed the throttles all the way forward to Takeoff/Go-Around thrust. Often simply called TOGA, it is maximum thrust, all the power the engines have to give, used on takeoff and when going around after a missed approach. Bonin must have finally realized they were in serious trouble.

At 11:06 and 6 seconds, one minute and one second after he took control of the plane, Bonin's airspeed indicator began working again. Now both pilots' PFDs were displaying perfect information.

What they saw next on their primary flight displays mystified them.

At just under 38,000 feet, with its nose pointed skyward and its engines in TOGA and operating full-out – at an N1 of nearly 100% – the big jet stalled and began descending. Since Bonin was holding his sidestick back, the nose remained pointed up as the plane fell. Their airspeed indicators showed – accurately – an airspeed of around 185 knots. They were moving forward and downward, much the way a plane moves in the landing flare.

Pilots have a name for this, descending in a stall (or close to it) with the nose pointed up: *mushing*. The big jet was mushing down.

Robert said, 'But we've got the engines. What's happening? [*Expletive*] Do you understand what's happening or not?' He could not understand why they weren't accelerating and holding altitude with full power pouring out of the back of his plane's turbines.

Nothing was adding up to them, and automation was offering no help. It was not stepping in to make everything alright again, or even just to tell them what was happening. ECAM was offering nothing useful. They may not have trusted their primary flight displays, but in the pitch-black sky they had nothing else to go on. If they had only taken a moment to concentrate on what their PFDs showed, they would have realized they were mushing and would have known what to do. But they could not get there.

They were in one of the most automated cockpits in the world, with all their instruments and displays working, and they had no idea what their jet was doing. They may have lost spatial orientation, and they had definitely lost their all-important situational awareness.

A few seconds after the plane began mushing, Bonin said, '[*Expletive*] I don't have control of the airplane any more now.'

But he was not giving up. He was still trying to exert his will over the plunging airliner. When the mushing began the jet had banked to the right. He countered by moving the sidestick left. But wings of a stalled plane don't properly cooperate with their ailerons. The A330 remained banked right, beginning what would eventually be a complete U-Turn while falling at around 10,000 feet per minute. At that rate-of-descent, the beleaguered pilots had two and a half minutes to solve the puzzle of what was happening to their craft. By then they would be at 10,000 feet altitude, too low to maneuver out of their jam.

One and a half minutes after this all started, Robert clicked a red button on his sidestick, and said, 'Controls to the left.' He was taking over.

That lasted one second. Bonin clicked his own red button and seized the plane back. He did it silently, without a word spoken, so Robert may not have initially realized he had lost control.

As soon as Bonin re-established authority over the jet, he pulled the sidestick full left, and full back.

He moved it left to counter the right bank, though it still had no effect.

He moved it back for reasons that are, and will likely always remain, a complete mystery. Pulling back lowers the tail, which raises the nose, or in this case, keeps it raised, which also slows the plane down. Bonin needed the opposite – he needed to gain airspeed, lower the angle-of-attack, and fly out of the stall. The only two ways to do that was to add power, or to lower the nose below the horizon so the jet would fly 'downhill.'

They were already at TOGA power. The obvious next move was to lower the nose.

Bonin did not.

Now Dubois returned to the cockpit. He needed to quickly understand what the jet and his pilots were doing. This was their first conversation.

Dubois: *Er, what are you doing?*
Robert: *What's happening? I don't know. I don't know what's happening.*
Bonin: *We're losing control of the airplane there.*
Robert: *We lost all control of the airplane. We don't understand anything. We've tried everything.*

No, they hadn't tried everything. They could have lowered the nose.

Better yet, Dubois should have replaced Robert, his relief pilot, in the captain's seat and reassumed command of his flight. But he did nothing of the sort. Instead, he sat in the jump seat between the two co-pilots. He had nothing to contribute to arresting the plane's descent. Ignoring it entirely, he urged Bonin to get the wings level and out of the right turn by using the rudder. Bonin still had the sidestick full left, but since the ailerons of stalled airplanes behave differently, every pilot knows – or *should* know – to use the rudder to level the wings.

Dubois's rudder suggestion may mean he realized the plane was stalled. But if he did, he never took the next step, demanding Bonin lower the nose.

They continued plummeting earthward, most of the time nose-high. It defies belief.

Finally, four minutes and 23 seconds since the autopilot disengaged, having turned 180 degrees around to a heading of due west, away from Europe, the recording ended as the plane belly-flopped into the Atlantic Ocean at a vertical speed of around 120 miles per hour and a forward speed of 107 knots.

The impact killed everyone aboard.

Chapter 44

No Excuse

Remarkably, the tiniest mechanical parts can lead to the loss of the largest airliners. We have read so far about a forgotten cockpit switch in one, a broken angle-of-attack vane in two others, and now temporarily clogged pitot tubes. Small, relatively inexpensive pieces of hardware were the cause of hundreds of deaths.

But these parts did not bring down the planes by themselves. They may have produced innacurate data or surprising results, but in the final analysis the pilots had control, and had options to deal with their respective troubles.

That is particularly true in the case of Air France 447. Something else is killing these experienced pilots and their passengers.

The immediate cause of 447's loss, the act of commission that directly led to the ultramodern Airbus A330 widebody crashing, is easy to identify: Bonin's nearly-continual nose-up input on his sidestick keeping the A330's tail down and its nose high as it mushed earthward for almost 38,000 feet.

The longer-running links in the chain of events that brought the airliner down are equally easy to spot. Two intertwined links are the faulty pitot tubes, and Air France management's failure to replace them. The airline knew they were problematic. They could not clear the ice crystals as fast as they were ingesting them, and the improved model should have been installed as soon as possible. These were not trivial parts like toilet seats or galley coffee makers. The delay was deadly. The airline should have hustled hard to get every plane in the fleet updated.

Another link in the crash is the performance of Air France's Operations Control Center. After takeoff the Center warned the flight crew of weather ahead through ACARS. But its warning came too late to keep Bonin from a severe case of nerves. It should have alerted the crew before they had even stepped into the cockpit, when they were flight-planning their route. They could have charted a course away from the storm cells. It is hard to believe the genesis of that storm wasn't percolating in the atmosphere a few hours earlier. The Control Center could have been more on top of things.

Connected to this cause is Captain Dubois' indifferent attitude towards the severe weather in his flight's path. He probably had no intention of flying through the storms Bonin spotted on radar, but his lack of concern, even if perhaps nothing more than his way of showing his superior rank to Bonin, did not help the junior pilot's mood and may have contributed to his anxiety, which eventually manifested itself in his keeping the sidestick back as they plunged earthward.

These causes, while important, are incidental. In the end this crash and loss of life comes down to the cockpit crew, to Bonin pulling back on the sidestick while his more experienced cockpit crewmates did nothing to stop him.

Such poor airmanship among the three pilots aboard that airliner has its own cause, a third link in the long-term causal chain – perhaps a link that dominates all the others. Once again that cause – that link – is automation.

While all three crewmembers displayed remarkable ineffectiveness, Bonin's behavior is the hardest to understand. We can guess why he raised the plane's nose high *at first*. Like Marvin Renslow on Colgan 3407, he may have been startled when the autothrottle and autopilot disengaged. The French government's BEA report confirms it: 'This degree of surprise can be explained by the contrast between the triggering of the warning and the situation in the cruise phase, during which the pace of change tends to be slow and concentration levels are lower.'

A phrase to describe what for pilots is often 'the situation in the cruise phase' is one we know well by now: automation complacency. Bonin had gone from the complacency-inducing, Zen-like calmness of high-altitude cruise to sudden crisis.

He initially handled it well. After declaring, 'I have the controls,' he first (probably) noticed his primary flight display's altimeter showing they were at 34,670 feet. As an aviation professional, his instinctive reaction to being shocked into the present should have been to return the jet to their assigned altitude of 35,000. Maybe that is what he was thinking when he first pulled back on his sidestick.

But once he got the plane up to 35,000 feet where it belonged, he continued pulling back. The question is, why?

The BEA accident investigators, as perplexed as anyone, asked that in their report: 'Although [Bonin's] initial excessive nose-up reaction may thus be fairly easily understood, the same is not true for the persistence of this input, which generated a significant vertical flight path deviation.' [*Vertical flight path deviation* means he climbed higher than he should have.]

Bonin's initial automation complacency was coupled with another Hazardous State of Awareness, Excessive Absorption. The copilot's attention was tunneled in on the storms ahead of them. With his drinking-straw focus on the weather, he was completely oblivious to, and complacent about, the plane's performance. Autoflight was in control, and Bonin, the Pilot Flying, wasn't giving autopilot and autothrottle a single thought. So, as the BEA suggests, he may have been startled when they disengaged. Coming out of the Complacency Hazardous State of Awareness, he was consumed by a seemingly panic-level fear of flying in those clouds. All he wanted was to climb above them, even though he had no way of knowing whether he could have climbed high enough to break into the clear.

His behavior did not change once the jet stalled and the nose-up mushing descent began. At that point Bonin had utterly no reason to hold the sidestick back. In retrospect, he did not react as Renslow did. He reacted *worse*. Both of them forgot everything they had ever learned about stalls, but Renslow's reaction lasted seconds; Bonin's lasted minutes.

And how about Robert, sitting next to Bonin, watching the copilot mishandle the plane. Why was he so lackadaisical, letting his crewmate wallow in ineffectiveness? After Bonin took over hand-flying the jet, Robert could not see what the copilot was doing with his sidestick. Unlike planes with yokes, Airbus sidesticks operate independently of each other. They are designed with the expectation that only one is used at a time. One pilot's inputs are not reflected in movement of the other pilot's sidestick.[67]

The only way Robert could have known what Bonin was doing with his sidestick was to look at it, or ask him. It appears he did not look at it, or surely he would have said something? And we know he didn't ask. It is also telling that he never moved his seat into proper position. It's like borrowing your much-taller friend's car and not adjusting the seat so you can reach the pedals and steering wheel. He seems to have remained complacent through the entire deadly episode, a Hazardous State of Awareness that may have begun during high-altitude cruise, but did not end until the plane hit the water.

---

67 If both pilots on a fly-by-wire Airbus use their sidesticks at the same time, the result is additive – computers add the two inputs and move the control surface accordingly, but in no case will it move a control surface more than the maximum that it would have moved with a single sidestick input. For example, if both pilots pull back 25% of the maximum, the result will be as if one pilot had pulled back 50%. But if they both pull back 75%, the result will be as if one pilot had pulled back 100%.

Robert was useless, except for his early order to divert left of their course to avoid the worst of the storms ahead of them. *Complacent* isn't the right descriptor for him. It is almost as if he remained in the BEA's *situation in the cruise phase* and never snapped out of it. It is mind-boggling, especially considering he had only spent eight minutes in the left seat before his comment to Bonin to divert.

Lastly, Captain Dubois's behavior is impossible to understand without attributing it to automation. First, as I said, his lack of concern with the weather may have been a big reason Bonin acted as he did. Then he left the cockpit knowing storm cells where just minutes away. A thunderstorm can rip apart the largest plane like it was made of matchsticks. No pilot ever knowingly flies through one.

Yet he walked out, showing not the least bit of concern.

He would only have done that if he had trusted the plane to get through whatever was ahead of them. The only source of that trust was automation. Even though Robert did the right thing diverting 12° left, Dubois would not have trusted Robert as much as he trusted Bernard Ziegler's automation and flight envelope protection.

When he returned to the cockpit, Dubois must have been startled by the unexpected picture greeting him – nose up and full throttle, but descending with the stall warning blaring and his pilots dumbfounded into inaction. It appears he could not initially comprehend the situation any more than the others. He might have been in the Mentally Fatigued Hazardous State of Awareness, given the weekend he spent in Rio with his girlfriend. He is reported to have slept very little during the layover, and being woken to an emergency was not the best way for his mind to come back to full-speed.

But he had more than two and a half minutes in the cockpit to get his bearings. With so much time, it is inexcusable that he had nothing concrete to contribute to the situation except to ask Bonin to level the wings with the rudder pedals. All his years as a pilot, and I would have thought he could have looked at the primary flight displays and said, 'Of course, I see. We're mushing. Lower the nose.'

Years of flying jets on autoflight had caused his flying skills to atrophy. Automation had crushed them.

So instead of taking command, the captain took the jump seat behind the two copilots and watched his plane descend into the South Atlantic.

Chapter 45

Hippocratic Oath

Automation did not simply lead to Complacency in the confused cockpit of Air France 447. It is the direct cause of the three pilots' inability to read and react to what they were seeing on their displays as their jet plunged downward.

An aircrew with nearly 20,000 hours in control of aircraft between them could not interpret their primary flight displays, instruments they had been using in one form or another their entire careers. They could not absorb accurate information about their flight projected in front of them and divine the condition of their airplane. They could not adjust for a sixty-one-second lapse in their airspeed aboard an otherwise healthy aircraft. Once they had lost situational awareness, it was gone forever.

Regardless of the roles automation complacency, automation bias (which we will see a few paragraphs ahead) and Mental Fatigue may have played, this positively screams out their lack of hands-on experience. The crew had spent so much of their professional lives on autopilot that when they really needed to fly, to be *pilots*, when they deperately needed their collective experience to save themselves, their skills had degraded to a vanishing point.

Here is one example. Had Bonin and Robert been up on basic knowledge of their jet they would have suspected a clogged pitot tube from the start. They would not have needed ECAM to point them in the right direction. The *magnitude* of the initial sudden altitude drop was the clue.

Information from the plane's pitot tubes is normally corrected for air pressure by a software algorithm in the plane's computers before being fed to the pilot's primary flight displays. That correction is approximately 300 feet. If the pitot tube is clogged and not working, that correction isn't necessary. When the pilots first saw the nearly 300-foot altitude change on their primary flight display, they should have immediately connected the drop with a pitot tube failure.

They didn't make the connection because of their lack of hands-on flying experience. Too much time monitoring displays and not enough time hand-flying and thinking about flying robbed them of their instincts and memory, of having once known this.

Pilots need to have their hands on the controls, not just on buttons and dials. If they don't hand-fly regularly they cannot be expected to react properly when things go south. It is asking too much of them.

Automation's role here was not just in making hand-flying a rare thing, and so diminishing cockpit crew skills. Its failure on Air France 447 began when it could not tell the crew what had gone wrong.

Early in the book, I said automation helps pilots by making it easier and safer to control their plane, and by keeping them up to date on the state of their aircraft. This is a clear case where automation failed to keep the crew informed and up to date. Their inability to read and interpret their instruments was exacerbated by automation's inability to concisely state what ailed their jet.

Since Robert and Bonin could not see out of the cockpit windows, their only clues to their jet's behavior, besides their bodies, was ECAM and their primary flight displays.

When Bonin took over and began hand-flying, Robert checked ECAM. It told him what he already knew – that autopilot and autothrottle had disengaged. It also said they had lost their flight envelope protections.

But it did not say why. It didn't say PITOT TUBE CLOGGED, or something along those lines. Robert was sitting in the captain's chair of one of the most computerized airplanes in history, and the computers could not tell him anything actionable.

But Robert still had options. For starters, he and Bonin might have consulted their own bodies. Their airspeed indicators showed they had slowed down from 275 knots to 60 in a few seconds. That is like smashing into a brick wall. At the same time, their altimeters showed they had dropped 330 feet, like falling from a 30-story building. We have talked about how the body can't be trusted when flying blind. But you *know* when you have instantaneously slowed by 215 miles per hour or when you are in a 330-foot freefall. The deceleration would have hurled them painfully against their seatbelts, while the fall would have rendered them temporarily weightless.

Bonin and Robert didn't feel any of that.

Like medical doctors' Hippocratic Oath, the first rule of piloting is, when something goes wrong, don't make it worse. The pilots failed to realize in those first few seconds that absolutely nothing had changed. They are not the first crew to encounter a faulty airspeed indicator while in cruise flight.

Safety experts have documented experiences of other aircrews successfully dealing with it. Those pilots handled the problem by doing nothing and in each case it turned out to be a non-event.

But not Bonin and Robert.

As for consulting their primary flight displays, the pilots believed they were wrong and not useable. Yet they never stopped reading them. Consider Bonin's initial hard-left input to level the wings, and then wing-rocking moves (if that is what he was doing). He could not see the horizon out of the windows yet he somehow knew the jet was banked. Same for Dubois, who upon returning to the cockpit also somehow knew the plane was banked.

They were both instinctively reading – and trusting – the two cockpit primary flight displays' attitude indicators (plus probably the backup indicator Robert had referenced to Bonin early in the crisis). Bonin's wing-rocking would have confirmed it. Yet at the same time, they were distrustful of their airspeed indicators and altimeters. They had no basis for making that distinction.

Within the primary flight display, the attitude indicator is predominantly how pilots maintain spatial orientation.[68] In the days before ADIRUs, mechanical/analogue instruments – altitude, airspeed and attitude – came in their own boxes and worked independently of each other. Airspeed and altitude readings were unaffected by an attitude indicator failure. But in modern planes using air data computers (the A-D in ADIRU), all the plane's sensors are plugged into a single avionics box. It is not unreasonable to think if one instrument on the PFD is impacted, they all are.

Yet all three pilots believed the attitude information. They were right to believe it, but their unquestioning faith in it while disbelieving everything else is perplexing and contradictory.

Once Bonin and Robert decided not to trust some of their instruments, the stall warnings only added to their uncertainty. As the plane stalled and the electronic 'STALL STALL' scratched at their ears, they must have been asking themselves, 'Are we really in a stall? Is this what a stall feels like? Is this what the primary flight display looks like during a stall?'

Or was the warning not to be believed?

They ignored it completely. They never even once mentioned it in conversation. After all, they were flying a jet they believed could not stall.

---

68 I say 'predominantly' because pilots maintain spatial orientation by using not only the attitude indicator, but the other instruments as well. In training for an instrument rating, pilots must be able to fly in clouds without their attitude indicator, so a proficient pilot takes in input from all the cockpit instruments.

Ziegler's flight envelope protection software guaranteed them that. Even though their primary flight displays and ECAM told them flight envelope protections were lost, they put that aside. It was preposterous that their A330 had stalled. They had never flown one in that condition. Even in training they had never stalled an A330. The last time they stalled a plane for real might have been years earlier, in primary flight training. After thousands of hours flying an essentially un-stallable plane, they could not believe it would be happening to them now.

That is automation bias at its most dangerous. Robert, Bonin, and Dubois were biased to believe the jet was not stalled, even though it had announced in bold letters in the ECAM that protections were off and that the jet could definitely stall. Just like the experiment where pilots accepted the less-than-optimal automated routings, here the pilots ignored stall warnings because they knew the Airbus would not let them down.

But it did.

Chapter 46

ECAM

ECAM, Electronic Centralized Aircraft Monitoring, is one of those advances pilots will one day wonder how they ever lived without. Its dual screens are found in the cockpit of every fly-by-wire Airbus.

The top screen, the Upper ECAM, displays engine performance information in its upper two-thirds and messages of general importance in the bottom one-third, such as whether the passenger cabin seat belt signs are lit or the plane's landing lights are on. The bottom third also shows warning and emergency messages when necessary, and complete checklists when something serious occurs, for example an engine failure on takeoff. During certain emergencies it even provides the helpful advice to LAND ASAP.

The Lower ECAM contains screens with greater detail about specific aircraft systems and components, such as the hydraulic or electrical system. Pilots can change the bottom display's contents at will, to monitor and troubleshoot in conjunction with the Upper ECAM screen.

Boeing has a similar system it calls EICAS, Engine Indicating and Crew Alerting System, though rather than providing entire checklists unprompted, it directs its pilots to a checklist they must either find manually on paper in the *Quick Reference Handbook*, or in electronic files they can retrieve on a different cockpit display.

ECAM's usefulness is readily apparent, but what if it can't put its electronic finger on the problem? That happened on Air France 447. For all the plane's wonderous sophistication, its ECAM's display did not help the beleaguered crew identify what had gone wrong. Instead it confused them, providing an array of symptoms that did not add up to a single simple truth: temporarily-failed pitot tubes. And one minute after the trouble began, when everything was working fine again, ECAM did not have the intelligence to say, 'All systems go, just fly the plane normally,' or something like that.

Taking it one step further, what if ECAM flashes a checklist on its screen in error, for a problem the plane is not having at that moment? Or what if the crew is not aware of anything having gone wrong, and therefore does not comprehend why a particular warning message and checklist have suddenly appeared? What do pilots do next in each of these cases?

Before addressing all the things that can go wrong with ECAM and EICAS, we should hear the other side of the story: times ECAM saved the day. One of those days it averted what would have been the worst single-plane disaster in aviation history. When an engine on a Qantas Airways Airbus A380 carrying 469 passengers and crew exploded soon after takeoff from Singapore on 4 November 2010, though the sheer number of ECAM messages nearly swamped the crew, it also helped them survive the immense damage to their plane.

The engine explosion had riddled Qantas flight 32 with 300 pieces of shrapnel, disabling electrical and hydraulic systems, degrading the performance of the three other engines, and starting a flash fire. The shrapnel holes in the wings impacted the plane's handling. But the highly experienced and oversized cockpit crew on the flight – it happened that five pilots were in the cockpit that day – spent two hours reading and working through every ECAM message, and while some of them were superfluous and unhelpful, most were not. The pilots brought the plane back to a safe landing and all 440 passengers and 29 crewmembers survived uninjured.[69]

ECAM proved its worth that day, even if it kept five pilots occupied for two hours reading and deciphering messages as they built a mental picture of the damage their jet had sustained from the engine explosion. But in spite of this success, ECAM is often rightly accused of not only providing useless information, but of inundating pilots with *too much* information. In an emergency, instead of focusing on flying the plane, two-man crews have found themselves drowning in a flood of scary messages and checklists that never once simply say, '*THIS* IS WHAT'S WRONG.'

Another near-disaster, coincidentally again involving a Qantas Airbus flying from Singapore, is a classic example of ECAM not being helpful. Automation aboard a Qantas A330 at cruise altitude mutinied. An inmate had taken over the asylum.

It was the scariest sort of automation surprise.

---

69 It was the passengers' good fortune that so many pilots were in the cockpit. The five included the basic long-haul flight crew of captain, copilot, and relief copilot. Also aboard was a senior 'check airman,' giving the captain a proficiency check, and an even more senior pilot checking on the checker. That is, not only was the captain undergoing a 'check ride,' but the check airman was in training, and so had his own instructor pilot watching over him. The crew totaled a remarkable 72,410 hours of piloting in their logbooks, more than eight *years* of total flight time. It is impossible to say if a smaller crew, or a less experienced crew, would have saved this plane.

Chapter 47

Seatbelts

Two years before the Qantas A380 engine explosion, another Qantas flight reminded the aviation community of the risks inherent in automation.

At 9:32 am on 7 October 2008, Qantas flight 72 departed Singapore's Changi Airport bound for Perth, Australia, five hours and 2,400 miles to the south.[70] Every seat on board was taken, totaling 303 passengers, nine flight attendants, and three pilots. In command was captain Kevin Sullivan, a 53-year-old California native and former US Navy F-14 fighter pilot who had married an Australian while on an exchange program with the Royal Australian Air Force. He had more than 13,500 total hours in cockpits.

Alongside him during takeoff was first officer Peter Lipsett, a former Royal Australian Navy helicopter pilot with 11,650 hours in his logbook. Accompanying the captain and first officer on this flight was a second officer, Ross Hales, aboard to give Sullivan and Lipsett the opportunity to take mid-flight rest breaks. Hales had 2,070 hours of flying time to date.

It was a perfect, cloudless and bump-less day for flying over the Indian Ocean and the Indonesian archipelago. Lunch was served when they were about half-way to Perth. By 12:30 the cabin crew had finished clearing the meal and had stowed the service carts. Captain Sullivan returned from a rest break at 12:33, and six minutes later the first officer, Lipsett, left the cockpit for his own time off. Second Officer Hales took the copilot seat and strapped himself in.

Cruising at 37,000 feet, at their 10 o'clock out of the cockpit windows Sullivan and Hales could make out the north-west corner of Australia 100 miles away, signifying they had around two hours left in the flight. Captain Sullivan was the Pilot Flying, the one handling autoflight settings and the one who would take control of the jet in the unlikely event the autopilot disconnected.

---

70 Times for this flight are in the Singapore time zone, which is the same as Perth, their destination.

THE DANGERS OF AUTOMATION IN AIRLINERS

At 12:40:28 that afternoon, 28 seconds after 12:40 Singapore/Perth time, that unlikely event happened: Autopilot #1 unexpectedly disengaged (the Airbus has two complete autopilots and uses one at a time).

Lightly gripping the sidestick controller in his left hand as he took over, Captain Sullivan, the ex-fighter pilot, waited a few beats as he calmly observed and assessed. Unlike the crew of Air France 447, he could see out of the windows into brilliant sunshine. His eyes were telling him nothing had changed.

ECAM displayed its first message: **AUTO FLT AP OFF**. AP – the autopilot – had disconnected itself.

This was obviously not news to Sullivan. Then more caution messages appeared, each accompanied by a doorbell-like chime. The messages did not have checklists associated with them and so did not require actions by the crew. But they were not helping Sullivan understand what was happening to his jet either.

Sullivan did not know that two seconds before Autopilot #1 disengaged, one of the plane's three ADIRUs began providing bad data to the fly-by-wire systems and flight computers. It had been working perfectly until that point. As designed, and as happened with Air France 447, in reaction to the bad data the autopilot had disconnected without warning.

Far below them the Indian Ocean appeared smooth and serene.

For the next twenty seconds Sullivan and Hales read a stream of ECAM caution messages, some new, others repeats. Then the intrusively loud '*STALL STALL*' synthetic voice alert went off in the cockpit, along with overspeed warning chimes declaring that the Airbus was flying too fast. The chimes did not have their own associated ECAM messages – Airbus designers considered the auditory warnings sufficient.

Those two alerts – stall and overspeed – were contradictory. A jet plane cannot be traveling both too fast and yet so slowly that it is about to stall. It made no sense to Sullivan and Hales either. But none of the messages scrolling by on ECAM shed light on what was actually happening. Trusting his gut, Sullivan elected to ignore the *STALL* voice and *overspeed* bells, and continued flying the plane as before.

Now among the cautions appearing on ECAM came the message, **NAV IR 1 FAULT**, declaring the jet's inertial reference navigation (NAV IR) system within the ADIRU was damaged. Sullivan saw a warning light illuminate on the panel above his head, confirming trouble with the navigation unit.

Yet that was not the biggest issue. A navigation unit fault would not have caused the autopilot to disengage. Maybe it was a new problem? Automation was muddying the picture.

While all this was going on, Captain Sullivan unintentionally let the plane drift upward 180 feet. He had been unconsciously pulling back ever-so-slightly on his sidestick (*slightly*, not at all like Bonin), a natural reaction when he had assumed control and turned his full attention to the ECAM warnings. As with Air France 447, flight controls are far more sensitive at high altitude and airspeed than when a plane is lower and slower.

Seeing his error, and interpreting the ECAM messages as implying nothing terrible was actually going on with the flight controls, he engaged Autopilot #2. With autopilot back in charge, he moved his hand away from the sidestick controller and watched the plane's computers gently guide them back down to 37,000 feet.

Forty-four seconds had gone by since Autopilot #1 had disconnected.

Fifteen ticks of the clock later, Sullivan noticed the airspeed and altitude readings on his primary flight display were fluctuating. Once again he saw nothing and felt nothing. As far as he could tell, the jet was flying straight and level. He glanced at the copilot's primary flight display and saw its instrument readings were steady, so he decided to ignore his own display and use Hales's instead. But, as required when airspeed and altitude readings become unreliable, Sullivan immediately disconnected Autopilot #2.

He was back to hand-flying the plane.

For most of the next minute, the two pilots continued reading ECAM messages as Sullivan expertly held the plane level at 37,000 feet. Nothing on ECAM was helping them understand the problem. Sullivan asked Hales to get Lipsett back into the cockpit. The captain intuitively felt something bad was brewing and he wanted the more experienced crewman in the copilot's seat. Hales got on the plane's cabin interphone to ask the Customer Service Manager in the forward galley (behind the Business Class seats in the front of the plane) to track down the first officer. Finding him would be easy, as at that moment Lipsett was standing in that galley.

Before the Customer Service Manager could pass along the copilot's message to Lipsett, the plane, on its own, seized control from Sullivan.

Chapter 48

Pan-Pan

With Sullivan smoothly flying the Airbus A330 at cruise speed and altitude without help from the autopilot, things seemed under control.

Then suddenly and without warning, the A330's huge elevators abruptly flexed downward into the slipstream, the tail rocketed upward, and the nose pitched down to an angle of 8.4° below the horizon. Accelerating earthward faster than the planet's gravitational pull, everyone aboard instantly went beyond weightless, to negative g's. Passengers wearing seatbelts felt blood rushing into their heads as their belts held them tightly. Those in the aisles or not belted in were hurled into the cabin ceiling, all of them hitting it simultaneously with a tremendous BANG. More than one-third of the plane's passengers and all the flight attendants were hurt in that instant, some of them seriously.

Sullivan's reaction was instantaneous and automatic: he pulled back on the sidestick to arrest the descent. For nearly two seconds the plane didn't respond. Then slowly it obeyed. Working skillfully and cautiously, in twenty-three seconds the captain had the nose back up to the horizon. The descent bottomed out 690 feet from where it started.

As Sullivan was gingerly guiding the plane back to 37,000 feet, Hales turned on the Fasten Seat Belts sign and keyed the public address system. 'All passengers and crew be seated and fasten seat belts immediately,' he ordered.

Ninety seconds after the sudden pitch-down Sullivan had the jet back at its assigned altitude. With Lipsett still in the main cabin – the Customer Service Manager had not gotten the chance to let him know he was wanted in the cockpit – Sullivan and Hales turned their attention to the ECAM, which they hoped would tell them why their jet had leapt out of their hands. ECAM had spat out four new messages:

NAV IR 1 FAULT
F/CTL PRIM 3 FAULT
NAV IR 1 FAULT
F/CTL PRIM 1 PITCH FAULT

The first two had checklists beneath them displaying required crew actions, which Sullivan and Hales executed quickly. The third was a repeat of the first, while the last had no action to go along with it. That took around one minute. Next the pilots queried the Lower ECAM screen, calling up diagrams and status reports on the plane's electronic and hydraulic systems, hoping to be clued into the jet's ailments.

Nothing was helping them understand what they were facing.

Then the plane came back at them.

One minute and ten seconds after returning to 37,000 feet, the elevators flexed down a second time. Again the tail shot up and the nose plunged down, though this time by a more moderate 3.5°. Again Sullivan pulled back on his sidestick, again the plane initially failed to respond – this time for almost three seconds – and then finally it cooperated. They lost 400 feet before Sullivan could begin climbing back to 37,000 feet again.

Now Sullivan himself got on the public address system, telling all passengers and crew to remain seated with their seatbelts fastened. Hales called the galley again asking for Lipsett, who had broken his nose in the upsets, but came forward and replaced the second officer in the copilot's seat. Hales took a cockpit jump seat and fastened his seatbelt tightly.

ECAM sent the pilots a new message: **F/CTL ALTN LAW (PROT LOST)**. The plane's protections were *lost*: the fly-by-wire system was now operating in Alternate Law, and the jet's flight envelope protections would not be protecting them. Additional ECAM messages alerted the pilots to primary computer and navigation inertial reference failures, same as before. Then more messages appeared. They were 'frequently scrolling,' according to the accident report by the ATSB, the Australian Air Transportation Safety Bureau, some of them repeating multiple times, and the pilots could not make them stop.

Meanwhile, also in the words of the ATSB, 'master chimes associated with the ECAM messages were frequently occurring, together with aural stall warnings and overspeed warnings.' Not surprisingly, during post-flight briefings the crew complained the cacophony of sounds and messages were a 'significant source of distraction.' In a jet where automation had taken over from the pilots, this sounds like they were soft-pedaling how bad it really was.

With all three pilots now in the cockpit and Sullivan at the controls, they debated what to do next – fly on to Perth, 700 miles to the south, or land at the nearest suitable airport. Given the undefined nature of the problems they were facing and the severity of the injuries to some of the passengers and flight attendants, landing as soon as feasible was the easy call. Because the plane's navigation software tracks alternative airports in case one is needed, they knew where to go: Learmonth Airport, a combination Royal Australian Air Force base and civilian airport less than ninety miles away to their east, the closest airport that could handle their widebody A330.

At 12:49 that afternoon, Lipsett, working the radios, issued a *pan-pan* call over their air traffic control frequency, saying Qantas 72 had 'flight control computer problems,' as well as injuries on board, and was requesting an immediate diversion to Learmonth.[71] The controller monitoring the flight quickly granted their request, ordering them to turn left towards the airport and begin their descent.

As Sullivan complied with the controller's directions, he asked Hales to call the flight attendants on the interphone and find out the extent of the injuries among the passengers and crew. When Hales reported that some of the passengers had severe injuries (including lacerations and broken bones), Lipsett rebroadcast the plane's distress call, this time declaring a *Mayday*.

Sullivan flew towards Learmonth cautiously, keeping their speed relatively low, their turn bank angles shallow, and holding his breath that the plane would not be hit by another pitch-down. At 1:32 that afternoon he landed the plane and rolled to a stop. The toll was 110 passengers and nine flight attendants injured, twelve of them seriously.

But their emergency was over.

---

71 Major in-flight problems are categorized by the FAA as one of two types: *urgency,* and the more severe, *distress*. The international radio call '*pan-pan*' (derived from the French *panne*, breakdown) is used to broadcast an *urgency* condition, defined by the FAA as 'a condition of being concerned about safety and requiring timely but not immediate assistance; a potential distress condition.' Mayday (from the French *m'aidez*, help me) is used when the safety of the vessel, passengers or pilots is at grave risk, and assistance is required immediately.

Chapter 49

Data Spike

Qantas 72 was two automation failures in one flight: sudden uncommanded pitch-downs, and atrociously poor ECAM messaging. The flight should be Exhibit A in a compendium of automation-related accidents, the first one anyone speaks of when discussing the topic. I suspect that it is not mostly because Captain Sullivan's flying skills, sharpened at the US Navy's TOPGUN air combat training school, saved the plane and kept it out of the public's consciousness.

But in this twenty-first century, when pilots and airline managements are relying on automation to both ease cockpit workloads and improve airplane safety, for this Airbus A330's automation to do neither is enlightening.

The pitch-downs are a textbook example of automation surprise. Nothing could be more surprising than the plane doing what it wants, when it wants. After it happened a second time, it left the crew in a state of extreme anxiety. Cedric Bonin on Air France 447 was overly afraid of weather along his plane's track. Captain Sullivan, on the other hand, would have had every reason to truly fear his plane's next outburst. He and his crewmates could not be sure it would not happen a third time, because they had no idea what caused the first two.

It would have helped if they could pinpoint the cause while still in the air. But they could not.

No string of events, no links in a chain, lined up to pitch Qantas 72's nose down at 37,000 feet. Instead the upset was caused by a complex and exceptionally rare series of incidents within the plane's fly-by-wire avionics that took place in less *than two seconds*. And it happened twice, back-to-back.

How rare was it?

This was the first time in 28 million A330 and Airbus A340 flight hours (one flight hour is one airplane in the air for one hour, and the A340 has the same avionics architecture that caused the problem on Qantas 72, so they are validly viewed together). That is almost 3,200 *years* of flight time.

THE DANGERS OF AUTOMATION IN AIRLINERS

What occurred inside the black boxes of Qantas 72's avionics shines a spotlight on the outer limits of automation. It seems some things could not be avoided by fly-by-wire's brilliant programmers, for example this one-in-28-million outcome. Though it is not wrong to blame the programmers for failing to account for it, the accident report by Australia's ATSB goes easy on them, writing, 'It is widely accepted that not all the potential failure modes and failure scenarios for complex systems can be identified in practice.'

Though the flying public would like programmers to have thought of every conceivable failure scenario, in the real world that is not possible. Some highly unlikely scenario will be missed, or will be considered of such extremely low odds as not worth planning for.

This is one they should have planned for. Here is what happened.

All programmers, regardless of what they are programming, need to account for the possibility of incorrect data coming into their computer. It is an unavoidable fact of digital life. Data can be incorrect because, as with the MCAS crashes, a sensor fails. Or it can be wrong because of a software bug, electromagnetic interference, a one-in-a-million glitch, a cosmic event (sunspots send charged particles hurtling through space that can interfere with electronic gear), or for no reason at all. Remember how Boeing never found the reason why its 737 radio altimeters were giving false *USABLE* readings.

One form of incorrect data is a *data spike*, a sudden burst of too-high numbers coming from a sensor, or going from one computer to another. For example, a spike might be found within a series of airspeed readings, each half-a-second after the other, that look like this:

310 knots. 310 knots. 309 knots. 310 knots. 450 knots. 310 knots. 310 knots…

The 450 knots can't possibly be right – it is a spike. Yet this sort of thing happens in avionics, and programmers must be sure a plane's computers can weed them out.

Angle-of-attack data is particularly sacred, since a plane flying at too high an angle-of-attack risks stalling. It is imperative to know quickly if angle-of-attack data that seems out of line is a spike or the real thing. Here is another example of a spike, this time within a series of angle-of-attack angles, each of them half-a-second apart:

4.0°… 4.0°… 4.1°… 4.0°… 4.2°… 4.1°… 4.1°… 50.6°…

Jets always cruise slightly nose-high, 4.0° being a typical nose-up angle.[72] Confirm that for yourself next time you are aboard a flight in cruise – walking back towards the tail feels a bit downhill, while walking forward seems slightly uphill. But is 50.6° a spike? Or is it the plane's new angle-of-attack? It would be impossible for a plane's nose to lurch up so steeply that rapidly, but how would the software know it hadn't actually happened?

One clue would be the next angle-of-attack number. In the example, if the next number in the sequence was 4.0° or 4.1°, then 50.6° was a spike. But if the next number was 50.6° again, the program would believe the plane had, in reality, lurched up that much.

As you know already, the ADIRU takes in data from sensors around the outside of the airplane and translates them into airspeed, altitude, heading and angle-of-attack information. Until now it wasn't necessary to provide more detail, but now let's dig a bit deeper.

In an Airbus, ADIRU data is passed on to a computer called the Flight Control Primary Computer, nicknamed PRIM, and from there to the pilots' primary flight displays. The A330 has three ADIRUs, each getting data from different sensors, and it has three PRIMs, each getting their data from the three ADIRUs. FIGURE 16 is a simple diagram of what one ADIRU-PRIM-PFD combination looks like.

Software programmers are all about rules. One of the rules they gave the PRIM concerns bad data. If one of the ADIRUs suddenly provided data that seemed wrong, the PRIMs were programmed to ignore it, but as a precaution, they would also disconnect the autopilot. That happened to Air France 447 and Qantas 72.

Because the angle-of-attack is so important, another programming rule specifically pertained to that data from the ADIRUs. If an angle-of-attack number was suspected of being a data spike, the PRIM was programmed to ignore new data for 1.2 seconds. After 1.2 seconds elapsed, the PRIM would use new angle-of-attack data.

What if a second, identical spike occurred exactly 1.2 seconds after the first spike? In that case the PRIM would see the two spikes are identical and think, *'It must not be a spike – the plane must really be tilted up at this crazy angle-of-attack.'* Flight Envelope Protection would then order the elevators down, raising the tail and shoving the nose downward.

---

72 The angle-of-attack during cruise could be lower, or higher, depending on the airplane type, its weight, altitude, cruising speed, and other factors impacting aircraft performance. 4° is a useful point for this example.

That simple description is what happened to Qantas 72. Two data spikes, 1.2 seconds apart, that the PRIM believed were real. Thinking the plane's angle-of-attack was alarmingly steep and close to a stall, the PRIM acted to save the plane.

Let's walk through the entire sequence, from the beginning.

At 11:40 and 26 seconds, Qantas 72's ADIRU 1 began throwing out erroneous data – airspeeds, altitudes, headings and angles-of-attack that had no bearing in reality. Seeing that, the PRIM disconnected the autopilot.

At 11:42 and 27 seconds, two angle-of-attack data spikes hit the PRIM 1.2 seconds apart. Believing they were real, the PRIM pushed the nose down and Captain Sullivan responded.

Almost three minutes later, another pair of data spikes came 1.2 seconds apart. Again the PRIM believed them and pitched the nose downward.

PRIM's software is not MCAS, but it is a first cousin. MCAS was a software band-aid applied to the 737 MAX because the aerodynamic forces created by its airframe and engine nacelle working together could cause it to pitch up into too high an angle-of-attack. On the A330, PRIM had software programmed to watch for a too-high angle-of-attack no matter how the plane got there.

Both the Boeing and Airbus software forced the nose down. But Boeing's MCAS could not let go until the planes crashed because it was reading a continual stream of false data from broken angle-of-attack sensors. PRIM's software could see, once the spike had passed, that it had been given bad data and allowed Sullivan to resume control.

The Australian ATSB accident report lauded Captain Sullivan for his responses to the uncommanded pitch-down elevator movements, saying this:

> The captain's sidestick responses to both pitch-downs were prompt… Given that the situation was sudden and unexpected, there was a risk that the flying pilot [Sullivan] would have overcorrected (that is, provided an excessive sidestick response), which would have led to more severe vertical accelerations during the recovery. However, in addition to being timely, the captain's sidestick responses were also of the appropriate magnitude.

Unlike the crew of Air France 447, *the situation in the cruise phase* did not lull Captain Sullivan into automation complacency. There is no substitute for experience and coolness under pressure.

Chapter 50

Distraction

ECAM was the other automation problem aboard Qantas 72. While it did not cause the sudden pitch-downs, its messages were of no value in correcting them either. Automation had failed in one of its most important cockpit missions, keeping pilots up to date. Worse, it was distracting to Sullivan and his copilots as they flew their damaged plane.

Distracted pilots can lose focus, and they can lose situational awareness. When that happens, accidents are not far behind. It is the reason for the Sterile Cockpit Rule below 10,000 feet.

Distraction has become closely associated with texting while driving. Though that is a perfect example, distraction is much broader than that. It is anything taking a person away from the primary task at hand.

While driving, it includes checking your nav and talking to your passenger.

While flying, it encompasses anything pulling pilots away from fully concentrating on their flight. That includes flight attendant calls, ACARS and ECAM messages, communicating with air traffic control, and looking for other airplane traffic. Even reviewing checklists, though vitally necessary, is distracting because when the pilot and copilot are focusing on whether a switch is set or a light is illuminated, they cannot be focusing on whether their plane is maintaining its assigned altitude and heading, or holding its airspeed.

An extreme example of the harm distraction can do was demonstrated in December 1972 by the pilots of Eastern Airlines flight 401, a Lockheed L-1011 flying from John F. Kennedy Airport in New York to Miami International. Around 11:30 pm, on a clear, pleasant but moonless night, after lowering the landing gear on the approach into Miami the crew noticed the nose gear's *down-and-locked* light on the instrument panel was not lit. Often when the light fails to illuminate, the gear is actually in position, so the crew needed to investigate further. Air traffic control instructed the plane to fly north at 2,000 feet so the pilots could figure out what had gone wrong.

The copilot, the Pilot Flying, turned on the autopilot and then pulled the suspect bulb out of the instrument panel to check if it had burned out. He could not tell, and he couldn't replace it in its socket, so the captain sent the flight engineer down into the avionics bay underneath the cockpit floor to see for himself if the landing gear was down and locked.[73] As it was pitch black in there, the flight engineer could not see a thing.

While they were all preoccupied with figuring out if the landing gear had deployed, no one realized autopilot was not maintaining their altitude (the accident report gives a number of possible reasons for the downward drift, none of them conclusive). The jet slowly descended to just a few feet above the Everglades. Suddenly the captain noticed they were too low, but it was too late. Of the 176 passengers and crewmembers aboard, 99 died in the crash.

Distraction can be deadly.

While Captain Sullivan, aboard Qantas 72, was struggling to identify the gremlins hijacking his jet, ECAM was barraging him with messages that were not pointing to the source of the trouble. From the moment the autopilot disconnected, Sullivan and his copilots encountered twenty-one separate ECAM caution messages, many appearing multiple times, and eight of which the ATSB report called 'spurious.' This total does not include the *overspeed* chime and *STALL STALL* warnings that blared and dinged in the crew's ears nearly continuously from the autopilot disconnection through landing.

The stress level in the cockpit must have been immense. Fearing their plane might again grab the sidestick out of their hands at any instant was at least disconcerting, and probably frightening. In understated fashion, Australia's ATSB acknowledged the negative impact ECAM was causing, though not the stress it was contributing, writing, 'The large number of spurious warnings and caution messages that resulted from the anomalous air data inertial reference unit [ADIRU] behavior created a significant amount of workload and distraction for the flight crew.'

It is a wonder Sullivan, as the pilot handling the plane, didn't fall into an excessively-absorbed Hazardous State of Awareness, so focused, so tunneled in on a single aspect of his flight that he missed other equally important signals about what his jet was doing. Falling back on experience,

---

73 L-1011s had a three-man crew of pilot and copilot, plus a flight engineer to monitor aircraft systems. This was typical of flight crews in the early days of long-range airliners (starting with piston-powered planes), before automation eliminated the need for the flight engineer position.

he probably turned the HSA to his advantage, shifting to a vigilant conscious state which Pope and Bogart in their 1992 report on HSAs called 'active and focused monitoring,' rapidly and intensely scanning his instruments for clues and warnings of what might come next. That behavior was a direct result of his years in the cockpit and his fighter pilot training. He had nothing specific to focus on, no single guilty instrument to stare at. All he had was the excruciating wait for the next pitch-down. But he stayed appropriately alert.

The copilots, first Hales and then Lipsett, may also have been in that hybrid vigilant state, taking in as much information as possible, searching for outliers that could kill them. They too might have raised their levels of awareness without sacrificing attention to the big picture.

A study by human factors experts at NASA's Ames Research Center has an interesting explanation for why distractions are so dangerous. In a very general summary, it suggests human functions are carried out by two cognitive systems, one automatic, the other conscious. Automatic functions require little or no thought. Cycling down a quiet bike path on a nice day, for instance. Conscious functions expend serious brainpower. Reading instructions for a new cell phone is an example.

The report's authors believe 'a task requiring a high degree of conscious processing … cannot be performed concurrently with other tasks without risking error.' That is a direct knock on multitasking. If you have ever tried having a conversation while concentrating on reading a news article, you know the feeling.

Chesley Sullenberger, the captain of US Airways flight 1549, that famously ditched in the Hudson River (a story we will cover in the next chapters), is on record saying that true multitasking is a myth. He suggests that when we think we are pulling it off we are actually switching rapidly back and forth between tasks, doing neither of them well.

Reading and interpreting ECAM messages when your plane is in trouble requires a high degree of 'conscious processing.' You are searching for clues to your craft's problem and you can't miss anything. On Qantas 72, ECAM's flow of useless messages, all emanating from the automatic heart of the Airbus A330, put the crew in greater danger than they already were.

Eventually the pilots got an ECAM message telling them ADIRU 1 had experienced a fault. It came 28 minutes after the second pitch-down, less than 20 minutes from landing. And even then, the crew had no idea the message was identifying the system causing the pitch-downs. It was merely a further distraction.

The ECAM aboard Qantas 72 had worked as designed. But engineers at Airbus had not anticipated a series of angle-of-attack data spikes perfectly placed 1.2 seconds apart. So they had never created a message for it. Instead, ECAM blurted out whatever its electronics told it to say, scaring the pilots and taking them off their primary task, which was to fly their plane and land it safely.

An obvious pair of questions arise when thinking about the travails of Qantas 72's pilots. First, if the software programmers who produced the plane's programmable systems are not responsible for what Sullivan's crew encountered, how can it be prevented next time?

And second, how can automation ever truly be completely trusted if it can go berserk without either cause or warning? How can you accept driving at highway speeds in an autonomously controlled car, or at 30,000 feet in a pilotless airplane, knowing what you now know about automation?

I think you should not.

Chapter 51

Canada Goose

The average Canada goose has a wingspan of nearly six feet and weighs between seven and ten pounds, females being a bit lighter than males. During migrations, they typically fly around forty miles per hour at an altitude of 2,000 to 3,000 feet, though they have been seen much higher.

A migrating Canada goose could do serious damage if it hit a small plane. If it hit the fuselage of an airliner it would leave a dent where it struck. One cold January, a flock of them did much worse than that. They caused a twin engine passenger jet to lose power in both engines, instantly turning it into a glider.

New York City's Central Park weather observatory on the afternoon of 15 January 2009 recorded a temperature of 20° Fahrenheit, a 10 mile an hour breeze out of the northwest, and a few clouds overhead. Locals would call it a perfect winter day for a walk. Pilots would call it a perfect day to fly.

A few miles east of the observatory, US Airways flight 1549, an Airbus A320, was slowing making its way through busy LaGuardia Airport taxiways to Runway 4. Sitting in the cockpit's left seat was 57-year-old Captain Chesley Sullenberger III, a former US Air Force fighter pilot who had been flying airliners since 1980, along the way accumulating nearly 20,000 hours in the air. His copilot that afternoon was 49-year-old Wisconsin native Jeff Skiles. He had been flying since he was 16 and now had 15,643 hours in cockpits.

When cleared by LaGuardia tower to take their 'position and hold' on Runway 4, Sullenberger, who had been taxiing the plane from the gate, lined up the Airbus with the runway's centerline and passed the controls to Skiles. The copilot would be the Pilot Flying. Then they waited for takeoff clearance while an airliner landed on a runway intersecting theirs.

Just before 3:25 pm LaGuardia Tower called them. 'Cactus fifteen forty nine runway four clear for takeoff.' 'Cactus' was the call sign used by all US Airways planes.[74] The other jet had landed and was out of their way.

Sullenberger, who had switched to Pilot Not Flying, acknowledged the tower. 'Cactus fifteen forty nine clear for takeoff.'

Skiles pushed the throttles up to TOGA power, and 32 seconds later he lifted the A320 off the runway. 'Gear up please,' he said, as the plane climbed away from LaGuardia into the sunny afternoon sky.

Sullenberger reached across the instrument panel and moved the landing gear handle to the UP position. Skiles banked the plane left and aimed straight north. As the gear tucked into their wheel wells, LaGuardia tower handed off US Airways 1549 to LaGuardia Departure Control.

Patrick Harten, 35 years old, Patty to his friends, was the air traffic controller working Departure. A triathlete and son of an air traffic controller, he cleared the flight to climb straight to 15,000 feet.

As the plane's flaps retracted, Sullenberger could not resist. 'What a view of the Hudson today,' he said.

Skiles peeked to his left. 'Yeah,' he agreed.

Little did they know.

Ninety seconds into the flight, at 219 knots and climbing through 2,700 feet over New York City's Bronx Zoo, Sullenberger suddenly said, 'Birds.' He had spotted a flock of them dead ahead.

One second after that, as the birds filled the windscreen like in a Hitchcock movie, the Cockpit Voice Recorder picked up sounds the NTSB report describes as 'thumps and thuds' – birds striking the plane. Later identified as Canada geese, at least one or more were sucked into each of the jet's engines.

On the CVR, the engine sounds decreased and Skiles said, 'Uh oh.'

Sullenberger quickly glanced at the ECAM's top two-thirds, displaying the engine information. He started to say, 'We got one roll–' and then corrected himself, 'both of 'em rolling back.' Their two engines were failing.

The captain went right to work from memory. 'Ignition, start,' he said, turning a switch to initiate relighting the engines.

Now pressing a button on the panel over his head, he said, 'I'm starting the APU.' The Auxiliary Power Unit, a small jet engine in the tail, could keep electricity flowing through the plane even without main engine power.

---

74 US Airways's 'Cactus' call sign in radio communication was inherited from America West Airlines, which had merged with US Airways in 2005. America West used 'Cactus,' a reference to its main hub of Phoenix, Arizona, to avoid confusion with 'American,' the long-ago-established American Airlines's call sign.

Then he took over the flying duties from Skiles with an authoritative, 'My aircraft.'

'Your aircraft,' Skiles said, removing his hands from the Airbus A320's sidestick and throttles.

Thirteen seconds had passed since the birds struck the plane.

Sullenberger now said, 'Get the QRH … loss of thrust on both engines.'

Skiles pulled out the *Quick Reference Handbook*. Sullenberger wanted him to read aloud the checklist for simultaneously losing power in both engines. It wasn't on the ECAM. Nothing worth looking at was on the ECAM.

Then Sullenberger radioed the words no pilot ever wants to say or hear: 'Mayday Mayday Mayday uh this is uh Cactus fifteen thirty nine hit birds, we've lost thrust in [or 'on'] both engines we're turning back towards LaGuardia.' He banked left to begin their return.

Harten, the Departure controller, replied immediately, 'Ok uh, you need to return to LaGuardia? Turn left heading of uh two two zero.'

'Two two zero,' Sullenberger repeated, confirming he understood Harten's instructions, and continued the left turn on his way to a southwesterly heading of 220°.

Using a different communications net, Harten cleared the airspace ahead of the damaged jet, ordering LaGuardia Tower to 'stop your departures, got emergency returning.'

In the cockpit, Skiles flipped open the QRH and began reading the Engine Dual Failure Checklist until he reached the third item down. He said, 'Airspeed optimum relight. Three hundred knots. We don't have that.'

'We don't,' Sullenberger echoed.

To restart – *relight* – its engines US Airways 1549 required air flowing into them at a speed of least 300 knots from the plane's forward motion. They were flying at 210 knots and slowing down, not close to what they needed.

Harten, the controller, radioed, 'Cactus fifteen twenty nine, if we can get it for you do you want to try to land runway one three?' Harten's call sign error, as well as Sullenberger's during his Mayday call, reflected the extreme pressure they were all under.

If Sullenberger continued his left turn he would end up pointing directly at Runway 13. But with the plane losing altitude at a high rate, he might run out of sky before reaching the runway's threshold. Still, Harten offered it.

Sullenberger had to assess quickly, and his decision would not get a second chance. He said, 'We're unable. We may end up in the Hudson.'

Sullenberger leveled the wings. He was now flying southwest. The Hudson river was a few hundred yards to his right and 1,600 feet beneath him.

Skiles continued working through the checklist. 'Emergency electrical power … emergency generator not online.'

Sullenberger. 'It's [or 'Is'] online.'

Harten, either not hearing Sullenberger's Hudson comment or not believing what he had heard, offered a second choice, the same runway but landing in the opposite direction. 'Arright Cactus fifteen forty nine it's gonna be left traffic for runway three one.'[75]

'Unable,' Sullenberger replied. He continued flying southwest. The plane was now over the eastern bank of the Hudson.

Harten said, 'OK, what do you need to land?'

Eight seconds of radio silence, then Harten returned, trying to be helpful. 'Cactus fifteen twenty nine runway four's available if you wanna make left traffic to runway four.' A call sign error again.

Right then US Airways 1549 passed over the New York side of the George Washington Bridge at 1,250 feet. The bridge's towers are 604 feet high. At their airspeed of 195 knots, the 646 foot gap between the tower and the plane was an alleyway. They didn't miss it by much.

Sullenberger replied. 'I'm not sure we can make any runway. Uh what's over to our right anything in New Jersey maybe Teterboro?' He could see Teterboro's runways out of the windows on Skiles's side of the cockpit.

Taking up the suggestion, Harten said, 'Ok yeah, off your right side is Teterboro airport. Do you wanna try and go to Teterboro?'

'Yes.'

Sullenberger tipped the wings slightly right, westward, angling towards the Hudson's New Jersey shore. Teterboro's runways were five and a half statute miles away.

Harten had previously worked air traffic control sectors in New Jersey and was familiar with Teterboro. He called the airport and asked if US Airways 1549 could land there. Teterboro's controller agreed immediately. Meanwhile Skiles continued working with Sullenberger and the QRH checklist to relight the engines.

With the Hudson river now 1,000 feet below them, Sullenberger's multiple responsibilities of flying the plane, communicating with Harten,

---

75 All runway pairs – the two opposing directions a plane can land on a single runway – are 180° apart. Runway 31 is the same runway but the opposite landing direction of Runway 13.

and responding to Skiles's checklist items was almost indescribably difficult to juggle. He made a decision even before Harten called back about Teterboro.

Two minutes after the birds struck, he switched his microphone to the passenger cabin public address system. 'This is the captain brace for impact.'

Immediately the flight attendants sitting in their cabin jump seats began shouting in unison, 'BRACE BRACE BRACE. HEADS DOWN STAY DOWN.' Over and over.

Harten. 'Cactus fifteen twenty nine turn right two eight zero you can land runway one at Teterboro.'

Sullenberger's response was definitive: 'We can't do it.'

Harten broadened Sullenberger's options. ''kay, which runway would you like at Teterboro?'

'We're gonna be in the Hudson.'

'I'm sorry say again Cactus?'

In testimony before the US Congress one month later, Harten said he was 'unable to wrap his mind around those words,' and needed to hear them again. He would not. That was the last exchange between Harten and Sullenberger.

US Airways 1549 was now 550 feet above the water.

As the plane neared the Hudson River, the cockpit's Ground Proximity Warning System piped up, alerting the crew to the approaching water with a loud, electronic, and non-stop, 'TOO LOW, TERRAIN. TOO LOW, TERRAIN,' and 'PULL UP. PULL UP.'

Ignoring the voice, Skiles gave his final summary about the engines: 'No relight.'

They were 340 feet above the Hudson.

Sullenberger prepared to ditch, saying to Skiles, 'OK let's go put the flaps out put the flaps out.'

Harten wasn't giving up. Even though the US Airways A320 was now too low to be picked up by air traffic control radar, he said, 'Cactus fifteen forty nine radar contact is lost you also got Newark Airport off your two o'clock in about seven miles.'

Sullenberger was too busy to reply.

They were now flying at 190 knots and 220 feet above the river, but needed to quickly get to 130 knots before the landing flare. Sullenberger raised the nose to fly 'uphill' and bleed airspeed.

Skiles said, 'Got flaps out,' as the Airbus rose back up to an altitude of 360 feet.

Skiles began giving altitude and airspeed call-outs so Sullenberger could focus on keeping the wings level. At their landing speed the river would be like concrete. If they hit with one wing even slightly lower than the other, they would cartwheel. If they hit with the nose not high enough they might plunge in and flip, or break apart. Either would kill everyone aboard. Sullenberger's focus on the aircraft needed to be total.

Skiles called their altitude. 'Two hundred fifty feet in the air.'

Then airspeed. 'One hundred and seventy knots.'

Ten seconds later, 'Hundred and fifty knots.'

He checked with Sullenberger about flap settings. 'Got flaps two, you want more?' Flaps-2 was 15°.

'No, let's stay at two.' Their airspeed was 135 knots.

Then Sullenberger asked his copilot, 'Got any ideas?'

Before Skiles could answer, Harten interjected with one last suggestion. 'Cactus fifteen twenty nine if you can uh…you got uh runway uh two nine available at Newark it'll be two o'clock and seven miles.'

No response.

Skiles may have spent those seconds thinking about what else they might do, but really, they were fully committed and out of options at this point. 'Actually not,' he eventually said.

Twenty seconds later – three and a half minutes since striking the birds – they hit the water at approximately 128 knots, nose high, wings perfectly level, and after a few hundred feet came to a stop. Sullenberger and Skiles, 35,000 hours in cockpits between them, both intuitively knew, from what they could see and feel and hear, that their Airbus was intact.

They looked at each other and at exactly the same time said, 'Well, that wasn't as bad as I thought.'

Chapter 52

Green Dot

It is not obvious where automation played a role in US Airways 1549, but in fact it did contribute, partly in ways that could have been harmful.

One thing is clear. Like Qantas 72, this accident has no links in a chain that led to the flock of Canada geese intersecting with the flight path of US Airways 1549, unless I consider global bird migration patterns. The story begins the instant Sullenberger recognized they could not dodge the flock of geese.

When Qantas 32's engine exploded, ECAM messages came fast and furious, at a daunting pace though they were ultimately mostly helpful. When Qantas 72's pitch-down gremlin attacked the plane, ECAM messages were a useless distraction. But for US Airways 1549, ECAM was nonexistent. It did not know specifically what had gone wrong, nor did it know how to fix what its sensors were seeing. All it knew was the engines were not performing properly. The NTSB's report on the accident said, '…currently no commercially available engines have diagnostic capabilities to identify the type of engine damage…and recommend mitigating or corrective actions to pilots.'

The engine damage could not be diagnosed by the plane's computers, so ECAM had no messages to display regarding next steps. After retrieving the engines from the Hudson and inspecting them, the NTSB determined neither had been capable of relighting, and no automation technology existed to make that call while the plane was airborne. It would have been helpful if there had been.

Instead, Skiles was forced to do what pilots had been doing for decades: consult a paper checklist. Like other accidents we have looked at, he and Sullenberger were in one of the most technically advanced aircraft in the world, and automation had nothing to offer them.

One place where automation provided help was in their airspeed. The plane's flight computers automatically calculated the airspeed US

Airways 1549 should fly to glide the furthest without power. It was called the *green dot* speed, because it was designated by a green circle on both pilot's primary flight displays' airspeed indicators. Sullenberger spent most of the time flying slightly slower than that speed, meaning the plane lost altitude faster than it might have, and it didn't glide as far as it ought to have.

The NTSB excused Sullenberger's too-low airspeed by raising the same distraction issues Qantas 72's crew faced with their unhelpful ECAM. The accident report said Sullenberger's list of simultaneous tasks was just too long, beginning with flying an engine-less plane, doing that while communicating with Harten, listening to the Ground Proximity Warning System yelling at him, seeing a huge bridge and tall buildings all around his field of vision, and following Skiles's checklist steps. It is a wonder he succeeded as well as he did.

The report cites the NASA Ames research report on cockpit distractions as proof of the difficulty facing Sullenberger. It concludes that the fact he flew a bit slower than he should have and glided a bit less far than he could have was simply being picky.

While we are speaking about gliding distance, an interesting question arises. If they had immediately turned back to LaGuardia within seconds of the bird strikes, and locked onto the green dot speed until the last possible second, would they have made it back to the airport? A feature film about the flight hypothesized that they would have.[76]

The NTSB tested the hypothesis in A320 simulators, and here's what its report said about it, with the underlined emphasis added:

> The simulations demonstrated that, to accomplish a successful flight to either airport [LaGuardia or Teterboro], the airplane would have to have been turned toward the airport <u>immediately</u> after the bird strike. The <u>immediate</u> turn did not reflect or

---

[76] The 2016 film, *Sully*, which is Captain Sullenberger's nickname, dramatically questioned whether Sullenberger made a mistake not heading back to LaGuardia. In the real world, the question was of minor interest because, first, everyone survived; second, Monday-morning quarterbacking Sullenberger's decision, given the stress he and Skiles were under, was just not fair; and third, it was clear from simulations – and for some pilots intuitively obvious – that turning back was the wrong choice. That is because pounded into every student pilot in their first days is that if their small plane loses an engine shortly after takeoff, never turn back to the airport you've just left – it is how pilots die (I am being simplistic here, but that is the gist of it).

account for real-world considerations, such as the time delay required to recognize the extent of the engine thrust loss and decide on a course of action.

If they had known instantly that the engines would never restart, would Sullenberger have turned back to LaGuardia right then? It's possible. But he didn't because he and Skiles did not know the extent of the damage. An NTSB simulation tried turning back thirty-five seconds after the bird strike and failed to reach a runway. Sullenberger's early decision to not try a return to LaGuardia was the right decision.

Though automation did not play much of an obvious role in US Airways 1549's accident, it could have done real damage during the last thirty-five feet of the ditching, during the landing flare as they neared the water. Sullenberger needed to raise his plane's nose just high enough and at precisely the right time so the plane would settle through ground effect relatively gently onto the water. If he did not get the angle high enough, the plane would not slow its rate-of-descent enough and it might snap in half when it contacted the river. On the other hand, if he raised it too high the jet might stall above the water and plummet nose-first into the Hudson, with who-knows-what result.

The Airbus A320 Sullenberger and Skiles were flying did not have engines, but it did have flight envelope protection. That had never gone away.

At approximately forty feet above the Hudson, when Sullenberger began to flare by pulling back on his sidestick to raise the nose, he intended to get the nose 11° or 12° up. Instead, the plane's flight envelope protection stopped him at 9.5°. It did that because the plane's rate-of-descent induced the software to think the angle-of-attack was high enough. Of course the whole point of raising the nose was to reduce the rate-of-descent – and it did – but it never got to where Sullenberger wanted it.

This is one of those situations Captain Rogers hypothesized about following his test flight comparison between an Airbus A330 and a Boeing 777 (discussed in Chapter 33). He noted the Airbus might prevent the pilot from achieving 'full aerodynamic capability as opposed to being software/control law limited.' Sullenberger found himself slightly software/control law limited. It is hard to complain about the eventual outcome, but it could have been a problem.

Sullenberger and Skiles in US Airways 1549, and Sullivan, Lipsett and Hales aboard Qantas 72, are perfect examples of the value of airmanship, of skilled pilots at the controls of wounded airliners.

THE DANGERS OF AUTOMATION IN AIRLINERS

Automation is great when it is great. But when something goes wrong, when a data spike appears, a radio altimeter fails, a pitot tube clogs, an ILS approach is not working, or a switch was set and forgotten, you want pilots who use automation to make their jobs easier, not ones who have allowed it to take over their professional lives.

Chapter 53

And Furthermore

I cannot repeat this often enough: the crashes described in these pages are the exception. Every man and woman within the global commercial pilot community comes to work each day expecting to achieve excellence. Some are better than others, some are luckier than others, but all try equally hard to get their passengers where they are going safely and on time.

In the Second World War, pilots needed 20/20 vision, an athlete's eye-hand coordination and perfect health before being allowed to train and earn a seat in a cockpit. Air-to-air combat in propeller planes was a blood sport and only the most gifted were allowed in the arena, because only they stood a chance of defeating the enemy's equally talented warriors. When the war ended many of those pilots came home to jobs with airlines, flying passengers across continents and oceans, stitching together again the world's fractured population.

As planes grew faster and passenger counts higher, automation in the form of flight management systems, autopilots, autothrottles, autoland, and flight envelope protection reduced the need for athletic excellence among pilots, and raised the requirement that they be vigilant, focused, alert and yet inactive for extended periods of time. Today when a crisis hits, stick-and-rudder skills are not often the deciding factor. Rather, it is the pilot's ability to remain cool and maintain situational awareness while reading warning messages and following checklists, pushing the right buttons and dialing the proper frequencies. That calls for a different skillset than the generals demanded of their pilots.

As we closed out the last century, human factors experts recognized commercial pilots' changing environment and began researching their ability to adapt to it. They asked questions:

Would a captain notice a gauge indicating a low-pressure-compressor problem if its light stayed green?

Would an auditory stall warning be heeded, or did the pilot require flashing lights as well, before recognizing the severity of the problem?

THE DANGERS OF AUTOMATION IN AIRLINERS

Would a pilot remember when autothrottle was programmed to automatically change modes, and notice when it did?

The answers were disconcerting. They discovered pilots aren't perfect. They don't always notice gauges, hear or see warnings, and remember settings. Then human factors researchers discovered the importance of eye movement. New questions arose:

Would pilots stare unblinkingly at their fuel gauge as it neared empty?

Would they stare at PAPI lights to the exclusion of everything else as they sank below a glide slope?

In an emergency would pilots widen their scan, taking in every instrument and quickly deciding what's important and what's unneeded? Or would they narrow their view to the proverbial drinking straw, staring at a single gauge while the world blew up around them?

Fredric Dehais, head of the Human Factors and Neuroergonomics department at ISAE-Supaero in Toulouse, France, a graduate-level aerospace school, says, 'Humans have the capability to deal with complexity, but the brain selects among pieces of information.'

How quickly and efficiently do human brains select those pieces? Do they select the right ones to monitor? Do they look at the right instruments but without really *seeing* and registering their contents? Or do they overload and shut out new data which might be crucial to their flight's survival?

Dehais is one of the world's leading experts on aviation human factors, continuing the work of Pope, Bogart, Parasuraman and others. He is now studying brain waves, seeing if pilots can self-monitor their own, perhaps connecting themselves to a cockpit light that warns them when their mental state is not what it needs to be, when it becomes Hazardous. He is wiring up cockpit crews to see if their brainwaves can mesh, if pilots can literally be on the same wavelength, and so improve the safety of their flight.

Boredom proneness is a thing. It is exactly what it sounds like: how susceptible is the pilot to becoming bored. Pilots who bore easily do not make good observers, which these days means they do not make good commercial pilots. Human factors experts are testing for it.

If we continue on this track, stick-and-rudder skills and airmanship will not be the primary determinants of whether a future pilot makes it in the airlines, or maybe even whether a pilot gets to the interview stage of a job search.

Even Boeing, standard-bearer of the pilot-first cockpit environment, is now hinting that it may alter its soft flight envelope protection philosophy, accepting more computer control. Its current CEO, David Calhoun, hinted as much when he said, 'We have always favored airplanes that required more pilot flying than maybe our competitors did. But we are all going to have to get our head around exactly what we want out of that. So that'll be a process that will go on alongside our next airplane development.'

That rethink is spurred by Boeing's continuing struggles with its 737 MAX. As of today the plane has been grounded for over a year, costing Boeing more than $19 billion. I am certain some within the ranks of Boeing senior management are wondering if the decision to produce the MAX, rather than building a new plane off a clean sheet of paper, was the right move. When the final cost is tallied, developing a direct competitor to the Airbus A320 may turn out to have been cheaper. This is one to watch.

We can hardly go five minutes in our everyday lives without automation interceding. It wakes us up, brews our coffee, and runs our commuter trains. It controls our cars. We can't make a restaurant decision without it. Come to work and find your company's internet access has been interrupted and you might as well go home for the day.

In aviation, automation enables you to buy your ticket and board your plane without ever interacting with another person. Air traffic control uses automation to keep planes on course and apart. Aboard the planes, automation does everything a pilot does, only better and faster – when it is working properly.

Forward-thinking manufacturers have publicly been considering that pilots may one day be unnecessary. That may already be true for *segments* of a flight today, and in the next century technology might be so perfect as to make pilotless passenger flights a reality. But I, for one, will never get aboard a plane that doesn't have a pilot. Wires fray, sensors fail, computer programs have bugs, birds migrate, and sometimes, as we now know, things go wrong for no reason at all.

I want a human sorting it out when I'm belted into my seat in the stratosphere. I want a human in the cockpit who can take control of my plane within milliseconds of a hiccup, or I am not taking my seat.

THE END

Acknowledgements

After the success of my first book, *Death March Escape: the Remarkable Story of a Man Who Twice Escaped the Nazi Holocaust*, I began considering a topic for a new book. The Second World War holds great interest for me, not the least because of my father's heroic survival of a concentration camp. So my attention was focused on finding a compelling story from that war that had not yet been told.

Then two Boeing 737 MAX airliners crashed four months apart. These planes were the newest and supposedly the safest jets in the sky, and the accidents dominated aviation news. As a pilot and unrepentant aviation enthusiast, I read all I could find on the topic, fascinated by the interplay between on-board computers, pilot skill, aerodynamics, and Boeing corporate might. Then my wife, Polly Leider, a journalist with a major broadcast network, pointed out that my computer engineering education, aviation knowledge, and finance background were the ideal combination to put the 737 MAX story on paper.

In researching the subject, I quickly discovered the MAX's issues were the tip of an automation iceberg. Automation in aviation, I learned, was a story with many sides, and it needed to be told.

So the first, and most important, person I must thank for helping make this book possible is my brilliant and beautiful wife, Polly. She not only encouraged me to begin the journey that eventually became this book, but her immense editing and wordsmithing skills came in handy often as my manuscript took form.

Dr. Alan Pope, an early pioneer in aviation human factors, lent me his time and knowledge as I worked to understand one of the most important issues confronting aviation professionals today. His invaluable assistance kept me on track as I absorbed the basics of this vast topic.

I tapped into the expertise of a number of past and present aviation professionals who asked to remain nameless, and so they shall. But still, among them there are three I must mention: the FAA employee, the former

ACKNOWLEDGEMENTS

fighter pilot, and the airline captain, who all read, edited and commented on book and chapter drafts. To you, thank you for your time and your input.

Finally, to the team at Pen & Sword, led by my extraordinary publisher Martin Mace, and in particular my seemingly clairvoyant editor Barnaby Blacker (he always seemed to know what I was trying to express) thank you for making all this possible.

Jack J. Hersch
New York, New York

Glossary

Many of the terms and phrases used within this book require a definition of more than a few sentences, but here I provide short explanations that hopefully give you enough to understand how I used them.

ACARS Aircraft Communications Addressing and Reporting System. A global data communications network linking cockpit crews with air traffic control (though it is not used to guide airplanes in flight), and with certain of that crew's airline departments on the ground, such as weather, operations, and maintenance.

ADIRU Air Data Inertial Reference Unit. It combines an Air Data Reference System and an Inertial Reference System within a single unit. The Air Data Reference system merges raw temperature, air pressure and angle-of-attack data taken from aircraft sensors to compute an aircraft's airspeed, altitude and angle-of-attack. The Inertial Reference system combines GPS, internal gyroscope, and accelerometer data to produce positional output such as attitude, ground speed, and flight path.

Ahead of the Airplane A term meaning the pilot is fully engaged with the plane and flight. A pilot 'ahead of the plane' is ready to turn, climb or descend when the route or air traffic control calls for it; has the right checklist ready at the appropriate time, and executes the items on that checklist on time and efficiently; is not rushing from task to task; and has good situational awareness.

Aileron A hinged control surface on the trailing edge of the wing, near the outer tip. It tilts up and down in response to commands from the autopilot, or the pilot's yolk, or joystick/sidestick. The upward or downward tilt changes the wing's camber, as well as deflecting air upward or downward, thereby causing the wing to fall or rise, respectively, which banks the plane. See Figures 9 and 10.

Air Traffic Control (ATC) The organization that monitors air traffic. It includes controllers in airport towers visually tracking aircraft on the ground and in the air, and controllers working solely with radar to track flights in the air flying through their area of responsibility.

Alternate Law see *Flight Envelope Protection*

Approach The final flight phase, from the end of the cruise phase through landing. The first segment of the approach is the Initial Approach, followed by the Final Approach. An Instrument Flight Rules approach has specifically defined points on the path to a runway delineating the Initial Approach and Final Approach, but the terms can also serve as approximations for the first half and second half of any approach to landing.

Approach Chart A page containing details of an Instrument Approach to a runway. An approach chart is in a standardized format and includes a map of the immediate vicinity of the airport, important frequencies (including those for air traffic control and navigation aids), airspeeds to be flown on a stabilized approach, rates-of-descent matching those airspeeds, missed approach procedures in case of the need to go around, and minimum flying altitudes to avoid the ground, nearby mountains and tall structures. Approach charts were once available only on paper and pilots were required to painstakingly keep up to date with the latest version of every approach. Now they are distributed and updated electronically.

ATC see *Air Traffic Control*

Attitude Indicator A stand-alone cockpit instrument or part of a primary flight display that graphically displays a plane's attitude in relation to the earth – nose up or down, and wings banked, and to what degree. See Figure 15.

ATSB Australian Transportation Safety Board. The government's accident investigation group in Australia, comparable to the NTSB in the US.

Autoflight see *Automatic Flight Control System*

Autoland An aircraft avionics system that merges autopilot, autothrottle, and internal navigation, to automatically track the glide slope and localizer radio beams of a runway's ILS Approach to land a plane. The aircraft's avionics and runway must both be certified as capable of executing such an approach with a sufficient degree precision.

Automatic Flight Control System (Autoflight) The combination of autopilot and autothrottle avionics systems. *Autoflight* generically refers to all the automatic flight control systems aboard a plane.

Autopilot An avionics system that automatically controls a plane's control surfaces. It follows directions either inputted by the pilot or from a Flight Management System to climb, descend, turn, hold altitude, and hold a specific heading. Along with autothrottle, it is part of a plane's autoflight system.

Autothrottle An avionics system that automatically controls a plane's throttles and therefore its engines. Autothrottle can be set to maintain a specific speed, or a specific amount of engine thrust.

Avionics A catch-all term for an airplane's on-board electronic components.

Bank The tilt of an airplane to the left or right. More technically, it is movement *around* its roll, or longitudinal, axis.

BEA The Bureau of Enquiry and Analysis for Civil Aviation Safety. The French government's aviation accident investigation organization, comparable to the NTSB in the US.

Blind Flying see *Flying Blind*

Buffet see *Stall Buffet*

Camber A wing's top and bottom surfaces are curved, with the difference between the two surfaces called a camber line. The top surface is usually more curved than the bottom (the bottom is often nearly flat), making the camber line itself curved. Camber is a measure of the curvature of the camber line.

Centerline A line running down the exact center of a runway. It is imagined extending many miles in front of a runway. Planes flying toward a runway line up with its 'extended centerline.'

Check Ride A flight where a pilot is tested for specific plane-handling skills by an examiner, who could be a government official, or an airline or flight school employee. The successful conclusion of a check ride could be, among other things, a new license, a new rating (such as an instrument rating), permission to fly a certain type of plane (for instance a twin-engine plane, or a 737), or permission to fly in a particular environment (for instance at night, or in mountainous regions).

Cockpit Voice Recorder (CVR) A self-contained device that records all the sounds within a cockpit. Its microphones can distinguish between conversation coming from the pilot, copilot, and jump seat occupants, as well as all ambient sounds. It is housed in a crash-proof container that can survive prolonged exposure to weather and water.

Commercial Aviation A catch-all term encompassing all aviation for-hire, including passenger and cargo airline operations, and small passenger jets not owned by an individual or a non-aviation-industry corporation.

Conga Line Slang term for a line of planes on a taxiway, usually leading to a runway.

Controller An individual working in air traffic control responsible for handling airplanes. A controller might direct plane movements on the ground at an airport, or in the sky using radar to identify and track the aircraft.

Control Surface The parts of an airplane that move, causing it to climb, descend, and turn. Primary control surfaces are the aileron, rudder, and elevator. Secondary control surfaces are flaps, slats, spoilers, and trim.

Convective Activity Essentially, thunderstorm activity. Comes from *convection*, the tendency of warm air to rise and cold air to sink. Warm air containing moisture is a key component of a convection cycle that produces thunderstorms.

Coordinated Turn A turn where the plane's nose is pointed directly in line with the turn's path, and the plane is not skidding or slipping. See *Skidding Turn* and *Slipping Turn*.

Cruise The long-range phase of a flight, where a plane is not climbing following takeoff, nor descending to the destination airport on the approach. For commercial airliners, the cruise phase of a flight is usually at high altitude, where jet engines operate most efficiently.

Cruise Speed (or Cruising Speed) The speed at which a plane can fly in the *cruise phase* of flight most economically (i.e., the furthest) with a full load of passengers.

CVR see *Cockpit Voice Recorder*

Dead-head Describes a flight-crewperson flying for free as a passenger on a flight (belonging to that person's, or a different, commercial air carrier) for the purpose of getting to or from an airport for work.

Decision Height During the final approach segment of landing, the height above the ground by which a pilot must decide to either land, or go around.

De-icing Removing ice from an airplane's fuselage, wings, tail, propellers and engines. On the ground it is done with heated liquid sprayed on the airplane through a high-pressure hose. In the air it is done with de-icing boots, heated liquid, or hot air blown onto the wing surfaces.

De-icing Boots Inflatable rubber bladders on the leading edge of a wing that are filled with air then suctioned empty in a rapid repetitive action, to break off ice that has accumulated over it on the wing.

Digital Flight Data Recorder (DFDR) see *Flight Data Recorder*

Direct Law or Mode see *Flight Envelope Protection*

Direct Linkage The use of cables and rods to connect control surfaces to the yoke, joystick/sidestick or rudder; as opposed to fly-by-wire, where electrical wiring connects the control surfaces to the cockpit.

Distance Measuring Equipment (DME) A cockpit radio receiver that measures the distance between a ground-based antenna and an aircraft. DME distances are used on landing approaches to airports to help pilots identify their location relative to the antenna, and therefore relative to the runway threshold. They are also used in navigation.

EASA European Union Aviation Safety Agency. The organization within the European Union responsible for civilian aviation, similar to the FAA in the US.

ECAM Electronic Centralized Aircraft Monitoring. Found in Airbus aircraft. Similar to EICAS on Boeing aircraft, it monitors and displays – on cockpit flat-panel screens – engine data and performance parameters, as well as the status of all aircraft systems. When anomalies and issues occur, ECAM either displays checklists of remedial steps, or if the checklist is lengthy, ECAM displays which *Quick Reference Handbook* checklist is to be used.

EICAS Engine Indicating and Crew Alerting System. Found on Boeing aircraft. Similar to ECAM on Airbus aircraft, it monitors and displays – on cockpit flat-panel screens – engine data and performance parameters, as well as the status of all aircraft systems. When anomalies and issues occur, EICAS indicates which *Quick Reference Handbook* checklist is to be used.

Elevator The control surface in the rear of the airplane attached to the horizontal stabilizer. When moved up or down, the elevator lowers or raises the plane's tail, thereby raising or lowering the nose and causing the plane to climb or descend. See Figures 9, 11 and 12.

FAA see *Federal Aviation Administration*

FCC see *Flight Control Computer*

Federal Aviation Administration (FAA) The US government agency in charge of all civil aviation and US civilian airspace.

Final Approach The last few miles of the approach phase to landing. At this point the airplane should be lined up with the runway centerline and descending along the glide slope. See also *Approach*.

Fixed Wing Aircraft Term referring to airplanes as opposed to helicopters (which are similarly referred to as *rotary wing aircraft*).

Flaps Secondary control surfaces that extend from the trailing edge of a wing. They emerge out and down, increasing the length of the wing's camber line as well as its camber. They raise the angle-of-attack and lower the airspeed at which the plane can fly without stalling. They are used on landing so a plane can fly slowly before touching down, and on takeoff so a plane can become airborne at a slower speed than if they were not deployed. See Figure 9.

Flare see *Landing Flare*

Flight Control Primary Computer (PRIM) In Airbus aircraft, a computer that links the ADIRUs, control surfaces, the pilot's primary flight displays (PFDs) and sidestick controllers. Flight data from the aircraft's sensors are relayed through the ADIRU and processed by the PRIM for display on both pilots' PFDs. The PRIM also takes the pilots' sidestick control movements and relays them to the appropriate control surfaces to move the plane as the pilot expects (subject to the limits of the plane's Flight Control Laws). See Figure 16.

Flight Control Laws see *Flight Envelope Protection*

Flight Data Recorder A self-contained device that records the positions of a plane's control surfaces, and many of its switches, dials and instruments, for the duration of a flight. Analyzed by accident investigators as they attempt to determine the cause of a crash. Like the Cockpit Voice Recorder, it is housed in a crash-proof container that can endure long exposure to weather and water.

Flight Envelope The complete range of maneuvers a plane can perform without stalling or experiencing structural failure.

Flight Envelope Protection A fly-by-wire aircraft with flight envelope protection uses software to give varying degrees of freedom to its pilots to handle the plane as they choose, while preventing it from stalling or exceeding the performance limits of the airframe. The degree of freedom is limited by specific Flight Control Laws. Airbus and Boeing each have three levels of 'laws' dictating the amount of protection a particular plane has. The law governing flight most of the time is Normal Law for Airbus aircraft and Normal Mode for Boeing planes. These provide the plane with full protection against the pilot banking too steeply, pitching up either too quickly or steeply, and flying too slowly, situations which, at their extreme, could cause a plane to stall. The next level is Alternate Law at Airbus and Secondary Mode at Boeing. These allow the pilot to stall the plane, with warnings as the plane nears the stall. The lowest level, Direct Law/Mode, has no protections and gives pilots direct control over the plane's control surfaces. Flight Control Laws operate in one of two basic philosophies: *hard protection* espoused by Airbus, or *soft protection* advocated by Boeing. In hard protection, the pilot cannot exceed the limits of the particular 'law' in control of the plane. In soft protection, as a plane nears the limits its controls become heavier and stiffer, but if enough force is applied a pilot can cause the plane to exceed the limits (to 'depart the flight envelope').

Flight Management System (FMS) An on-board computer in which a plane's entire route can be input by the flight crew, and the FMS will guide the plane along that route. The FMS can issue routing directions to the pilot, or it can instruct autoflight to follow the route.

Flight Plan A document (on paper or computer) detailing key data necessary for a flight, including departure and destination airports, distances between route waypoints, speed, altitude and projected elapsed times for each flight segment, fuel and fuel reserve requirements, winds and weather along the route, frequencies needed for communication with air traffic control and for navigation aids, and the plane's weight and balance (where passengers are sitting and luggage is stowed, to be sure weight is distributed around the plane evenly). Pilots must formally file a flight plan with the FAA if they are flying under Instrument Flight Rules. Otherwise they are not required to file one, though filing is strongly recommended.

Fly-By-Wire (FBW) Describes an airplane whose control surfaces are connected to the cockpit mostly through wiring and electrical connections, rather than through direct linkage with cables and rods.

Flying Blind Flying a plane without reference to the ground or horizon. Flying only on instruments.

FMS see *Flight Management System*

General Aviation The universe of owning and operating private planes. The International Civil Aviation Organization, a United Nations organization, defines it as anything not found within *commercial aviation* or using an airplane for work.

Glass Cockpit A cockpit that contains all flat-panel screens, rather than analogue gauges, dials and instruments. The flat-panels contain images, many of them digital versions of formerly analogue instruments, while others are near-duplicates of analogue gauges. Glass cockpits usually contain a version of a primary flight display (PFD), the six-in-one instrument that replaces the six-pack set of analogue instruments. See *Primary Flight Display*, *Six-pack*.

Glide Slope During the final approach to landing, the optimal path a plane should follow to the runway from 5-10 miles away from the threshold and a few thousand feet of altitude. It is usually a 3° upward slope measured from the runway touchdown point, though it could be slightly steeper if obstacles on the ground require a plane to remain higher for longer during the final approach. An ILS approach will have a radio beam visible on certain cockpit instruments that graphically displays the glide slope. For visual approaches and landings under Visual Flight Rules, runways often have PAPI or VASI lights to guide pilots, letting them know if they are above, below, or on the glide slope.

Go Around Describes a pilot's actions if a landing approach is aborted and a new approach needs to be established.

Ground Effect On landing, when the plane is within half its wingspan's distance above the ground, aerodynamic drag lessens while *lift* increases, together giving the pilot the sensation of a cushion of air having formed under the plane. Planes seemingly float in ground effect until settling down onto the runway, a phenomenon whose length depends on the airplane's speed.

Ground Proximity Warning System (GPWS) A cockpit warning system alerting the pilots that the ground is near. The warning is usually through a synthetic voice.

Gyrocompass A compass that uses a gyroscope to hold north, rather than using the magnetic attraction of north.

Gyropilot The commercial brand name for the Sperry Corporation's early-model autopilots.

Gyroscope A rapidly spinning wheel within a cage whose spin-momentum (technically *angular momentum*) holds its position in space. It is the core device within position-oriented instruments on aircraft, spacecraft, ships, and submarines, to give their operators (pilots and captains) information about attitude and direction. Digital flight instruments and air data computers use ring laser gyroscopes, which use sensors that detect rotational movement against lasers, rather than against a mechanical gyroscope.

Hand-flying Describes a pilot when flying a plane with hands on the yoke or joystick/sidestick and feet on the rudders. As opposed to an autopilot or autoflight controlling a plane.

Hard Protection see *Flight Envelope Protection*

Hazardous States of Awareness (HSA) Human conscious states first identified and described in research published in 1992 by Alan Pope and Edward Bogart. In *Hazardous States*, humans are either not as aware or vigilant as they ideally should be, or they are *Excessively Absorbed* by a particular object to the exclusion of everything else. It is a key factor in understanding automation's impact on pilots.

Heading The direction of flight in which a plane is pointed. Because of winds, the heading need not be the same as the *course*. For example, a plane heading north but flying in strong winds from the west (from the pilot's 9 o'clock), is actually flying on a north-easterly course (approximately in the direction of the pilot's 1 o'clock or 2 o'clock).

Horizontal stabilizer The wing-like structure in a plane's tail that gives it longitudinal (up-and-down) stability. The elevator is attached to the rear of the horizontal stabilizer. See Figures 9, 11 and 12.

HSA see *Hazardous States of Awareness*

Human factors The area of psychology covering human-machine interaction. Its study is intended to understand and improve how humans sense, use, react to, and interact with, machines and automation.

Icing The covering over of an aircraft's fuselage, wings, and/or tail with ice.

Identifier A 3-letter distinguishing code given to every airport by the International Air Transport Association. Used in communication, navigation and baggage handling.

Idle The lowest engine power setting, similar to Park or Neutral in a car. The engine continues running but produces minimal thrust.

IFR see *Instrument Flight Rules*

ILS Approach An approach using both a glide slope beam and a localizer beam to guide the pilot to the runway, enabling the pilot to reach the runway through clouds. Each ILS approach is described in great detail in its unique approach chart. See *Instrument Landing System.*

Indicated Airspeed A plane's speed shown on its cockpit's airspeed indicator. It is the speed measured directly by the pitot tube, and not adjusted for air pressure (essentially the plane's altitude), or air temperature. For any given Indicated Airspeed, at higher altitudes and higher temperatures a plane's True Airspeed – the actual speed it is moving through the air – is faster than at lower altitudes and colder temperatures.

Inertial Navigation System (INS) An avionics system that uses no outside information to compute and track where an aircraft is at all times. It works by being given a starting point, and then using sensors attached to gyroscopes and accelerometers, usually in conjunction with aircraft airspeed and altitude sensors, to continually track and update the aircraft's movement in three-dimensional space.

Initial Approach The first segment of a landing approach. Approach charts have a formal location for the beginning of the Initial Approach. However, less formally, the Initial Approach can refer to any point after the plane leaves the cruise phase of a flight.

INS see *Inertial Navigation System*

Instrument Approach An approach to landing where the pilot uses a plane's instruments to track a flight path to the runway. It is usually conducted under IFR (Instrument Flight Rules) conditions, though during VFR (Visual Flight Rules) conditions a pilot can conduct an instrument approach as long as he is also watching for other aircraft, or if a second pilot is aboard the plane to watch for conflicting air traffic. See *ILS Approach* and *Localizer Approach.*

Instrument Flight Rules (IFR) The conditions for flying where a pilot does not use visual references and is in continual contact with air traffic control. In certain weather conditions a pilot *must* fly under IFR. Those conditions include the cloud base below a minimum height, and visibility (the distance forward a pilot can see) less than a minimum. Planes must also fly under IFR when above 18,000 feet altitude. A pilot can always choose to fly under IFR regardless of weather conditions.

Instrument Landing System (ILS) A system of radio beams that an airplane's avionics can track to guide a pilot (or an autopilot or autoland) the last 5-10 miles down to a runway. An ILS uses two radio beams.

One provides the pilot with a glide slope (the *glide slope* radio beam) to follow. The other guides the pilot to line up the runway centerline (the *localizer* radio beam). See Figures 13 and 14.

Instrument Rating A rating a licensed pilot can obtain attesting to the pilot's ability and legal permission to fly in IFR conditions.

Instruments, flying on A term referring to flying without looking outside the cockpit windows, using only instruments for navigation guidance and to maintain a plane's proper flight attitude.

Intertropical Convergence Zone (ITCZ) An area near the equator where weather induced by the northern and southern hemispheres of the earth meet. It is characterized by both calm winds at sea level (known as the *doldrums*), and severe convective activity.

Joystick (stick) A stick attached to a cockpit floor, between a pilot's legs, that is attached to the elevator and ailerons, enabling a pilot to bank and pitch a plane. Pulling back on the stick makes the plane climb, pushing forward makes it descend, and moving it left and right makes the plane bank left and right. In a normally configured cockpit with a joystick, a pilot holds the stick with the right hand and controls the throttle with the left.

Jump Seat A fold-down seat in the cockpit that can be occupied by an observer, examiner, dead-heading pilot, or back-up pilot.

KNKT Komite Nasional Keselamatan Transportasi. The Indonesian transportation authority responsible for accident investigation, similar to the NTSB in the US.

Knot One nautical mile per hour. Its name comes from how sailing ships measured their speed as long ago as the 1600s. A piece of wood with a rope attached was thrown overboard. The rope had evenly spaced knots. The wood floated in place as the ship sailed away, with a sailor paying out the rope and counting the knots. The knots were about 47 feet apart and the count over 28 seconds was the ship's speed in nautical miles per hour.

Landing Flare The act of a pilot raising a plane's nose when a few feet above the runway during the landing segment of a flight. That action slows a plane, increases its wings' angle-of-attack and reduces its rate-of-descent, allowing it to settle into ground effect, which it will fly through until reaching the runway. Large commercial jets generally flare 30-40 feet above the runway, while small single-engine propeller planes flare approximately 10 feet above the runway.

Leading Edge The front edge of a wing or horizontal or vertical stabilizer.

LIFUS Line Flying Under Supervision. A final phase of flight training for commercial pilots. It entails flying regularly-scheduled passenger trips ('line flying') with an instructor pilot, or at least with a back-up pilot, to both observe, as well as to take over in case of issues. Not all airlines have a LIFUS training stage.

Localizer A radio beam identifying the runway centerline. It extends a few miles out from the threshold of a runway and is used by a pilot, autopilot, or autoland to hold the center line during an ILS Approach or Localizer Approach.

Localizer Approach An instrument approach that uses only the localizer radio beam, and not the glide slope beam. See *ILS Approach* and *Instrument Landing System*.

Mayday A radio call made when a pilot is facing an in-flight emergency categorized as *distress*, as opposed to a less severe problem categorized as *urgency*. Derived from the French *m'aidez* (help me), it is broadcast when the safety of the aircraft, passengers or pilots is at grave risk, and assistance is required immediately. [These categorizations and definitions are from the FAA website.]

MCAS Maneuvering Characteristics Augmentation System. A Boeing software suite produced for the 737 MAX aircraft to compensate, via software, for a design problem. When the plane is at high angles-of-attack – typically either nose up or in a steep turn – the aerodynamic forces between the plane's engine nacelles and fuselage may cause the angle-of-attack to increase on its own, bringing the plane close to a stall, or stalling it if already close. To compensate for this tendency, Boeing produced software to automatically detect these situations and reduce stall risk by inserting nose-down trim (which is done by deflecting the horizontal stabilizer upward). It is programmed to operate only when autopilot is not engaged and flaps are in the stowed position. If it malfunctioned, Boeing programmed it to appear to the pilot as runaway trim. See *Trim, Runaway Trim*.

MCP see *Mode Control Panel*

Missed Approach A failed approach to the runway. Also the name of the procedure executed when a plane must go around. A plane might execute a *missed approach* if another airplane has failed to clear the runway, or if its own approach is too high, or not stabilized.

Mode Confusion A description of pilots in an aircraft with too many choices of modes for autothrottle and autopilot. Pilots can make serious mistakes in setting and using modes, and can forget what modes do or how modes work with each other.

Mode Control Panel (MCP) On Boeing aircraft, the part of the instrument panel containing the buttons, knobs and dials for controlling the autopilot. It is just below the glareshield and between the two pilots.

Mushing (to mush) An aerodynamic condition where the plane is in a nose-high attitude but descending. The controls feel 'mushy' or sluggish during this sort of descent. A plane may be stalled, partially stalled, or not stalled at all when mushing.

N1 An engine performance parameter used by pilots to identify the amount of power coming from jet engines. It is the percent of maximum rotation speed of the low-pressure compressor. For example, 75% N1 signifies the low-pressure compressor is rotating at 75% of its maximum rotational rate if the jet were producing full power.

NASA National Aeronautics and Space Administration. The US government agency responsible for all civilian activities in space and space-related research.

National Transportation Safety Board (NTSB) The US government agency responsible for investigating all civilian transportation accidents, including car and highway, rail, sea and aviation accidents.

Nautical Mile Exactly 1.151 statute miles, or 1,852 meters. See *Knot*

Neuroergonomics The study of brain functions and mechanisms, and how those relate to the world of mechanization and automation. A field pioneered by Raja Parasuraman and today led by, among others, Fredric Dehais, current head of the neuroergonomics and human factors department at ISAE-SupAero, the French national aeronautics and aerospace institute of higher education.

Non-Normal Checklist (NNC) A checklist used when a not-normal event occurs during a flight. Typically the NNC will begin with memory items, actions that must be memorized by pilots and instantly recalled as soon as they recognize the application of a particular checklist is necessary.

Normal Law or Mode see *Flight Envelope Protection*

NTSB see *National Transportation Safety Board*

GLOSSARY

Pan-Pan A radio call made when a pilot is facing an in-flight emergency categorized as *urgency*, as opposed to a more severe problem categorized as *distress*. Derived from the French *panne* (breakdown), it is broadcast when vessels, passengers or pilots are facing a 'condition of being concerned about their safety and requiring timely but not immediate assistance; a potential distress condition.' [These categorizations and definitions are from the FAA website]

PAPI Precision Approach Path Indicator lights. A horizontal bank of four lights located just to the left of a runway, near the threshold, visible to a pilot on final approach, that change color based on whether a plane is above, on, or below the glide slope. The light sequence is: [White-White-White-White] = much too high; [White-White-White-Red] = too high; [White-White-Red-Red] = on the glide slope; [White-Red-Red-Red] = too low; [Red-Red-Red-Red] = much too low. See *VASI*.

PFD see *Primary Flight Display*

Pilot Flying The pilot responsible for controlling the airplane. In two-person commercial aircraft cockpits, normally the Pilot Flying physically flies the airplane (whether through hand-flying or through adjusting autoflight settings), while the Pilot Not Flying, also called the Pilot Monitoring, works radios, raises and lowers flaps, slats, and landing gear, communicates with the cabin crew, and monitors both the Pilot Flying as well as all cockpit displays and instruments.

Pilot Monitoring see *Pilot Flying.*

Pilot Not Flying see *Pilot Flying.*

Pitot Tube A pistol-shaped device on the outside of a plane that measures pressure created by its movement through the air. An Airspeed Indicator translates that pressure reading into Indicated Airspeed.

Primary flight display (PFD) A flat screen image containing all the flight information that was previously available only on separate, analogue flight instruments. The image presents a combination of digital and analogue data. Generally the six main cockpit instruments (commonly referred to as the *six-pack*) are shown on a PFD. They are: airspeed indicator, altimeter (for altitude), attitude indicator (the plane's attitude relative to the earth), heading indicator (compass direction), vertical speed indicator (rate-of-climb and rate-of-descent), and the turn-and-bank indicator (bank-angle of the wings, and whether the plane is slipping or skidding in a turn). See *Glass Cockpit* and *Six-pack*, and Figures 15, 17 and 18.

Quick Reference Handbook **(QRH)** A spiral-bound book of checklists and important information within easy reach of the pilots. In Airbus A320 models and newer, it augments information available on ECAM. In Boeing and other airplanes it is the primary source for checklists (including NNCs) during flight anomalies and emergencies. It also includes normal-operations checklists.

Radar Altimeter see *Radio Altimeter*

Radio Altimeter A device that computes the aircraft's altitude by the same principle as radar, sending a radio signal downward and measuring the time for the signal to return to a receiving antenna. It is extremely accurate below 2,500 feet above the ground and used by inertial guidance and by autoland during approach and landing.

Roll In flight, a tilting of the wings to the left and right. Synonymous with *bank*. In aerobatics, a roll is a 360° rotation of the aircraft around its longitudinal axis (the axis running through the center of the fuselage).

Rotary Wing Aircraft A term used for helicopters, as opposed to *fixed wing aircraft*, meaning airplanes.

Runaway Trim Situation when the electric trim mechanism in an aircraft (usually elevator trim) operates without instruction and on its own, causing the trimmed surface to deflect fully in one direction or another.

SA see *Situational Awareness*

SD see *Spatial Disorientation*

Secondary Mode See *Flight Envelope Protection*

Servo A small motor attached to a control surface.

Sidestick (or Sidestick Controller) A joystick a few inches in length and located on a pilot's side (left side for the left-seat pilot, right side for the right-seater). Operates the same as a floor-mounted joystick.

Situational Awareness In aviation, the concept of being fully aware of, and engaged in, everything pertaining to one's flight. This includes knowing where one is (attitude, altitude, airspeed), where one is going (the route, distances, headings, and obstacles along the way), communication and navigation frequencies to be needed (or where to find them), navigation aids to be followed, fuel quantity and flow rate, awareness of other planes in the sky, preparedness for emergencies, and everything else that comprises knowledge of one's entire situation.

GLOSSARY

Six-pack Slang term for the six primary instruments needed to control an airplane (not including engine instruments). They are: airspeed indicator, altimeter (for altitude), attitude indicator (the plane's attitude relative to the earth), heading indicator (compass direction), vertical speed indicator (rate-of-climb and rate-of-descent), and the turn-and-bank indicator (bank-angle of the wings, and whether the plane is slipping or skidding in a turn). See *Primary Flight Display*.

Skidding (turn) When in a banked turn, the plane's tail is outside the arc of the turn, appearing as a car might with its rear wheels sliding outside the arc of a turn.

Slats Secondary control surfaces that extend out and down from the leading edge of a wing. When they emerge, they widen the wing, lengthening the camber line and increasing the camber of the wing. They raise the angle-of-attack and lower the airspeed at which a plane can fly without stalling. They are used on landing so a plane can fly slowly, and on takeoff so a plane can become airborne at a slower speed. See Figure 9.

Slipping (turn) When in a banked turn, the plane's tail is inside the arc of the turn, appearing as a car might on a steeply banked roadway if its rear wheels were slipping down the bank and lower than the front wheels. Similar to understeer in a car.

Soft Protection see *Flight Envelope Protection*

Spatial Disorientation A loss of the knowledge of one's (or one's airplane's) attitude in relation to the earth. Can be thought of a losing reference to 'which way is up.'

Speed-on-Elevator Refers to autoflight controlling the airspeed of an airplane by using the elevator. Raising the elevator raises the nose and causes the plane to slow as it flies 'uphill.' Lowering the elevator lowers the nose and causes the plane to accelerate as it flies 'downhill.'

Speed-on-Throttle Refers to autoflight controlling the speed of an airplane by increasing or decreasing thrust via throttle movement.

Stabilized Approach A landing approach where few, if any, throttle changes and control surface movements are necessary to maintain descending flight at the proper speed along a glide slope.

Stabilizer (see *Vertical Stabilizer* or *Horizontal Stabilizer*)

Stall The complete loss of *lift* by a wing. It is the result of flying at too high an angle-of-attack. The critical angle-of-attack is the angle-of-attack when the stall occurs.

Stall Buffet Vibration of the aircraft as it nears a stall. It is caused by turbulent air created by the wings as their angle-of-attack nears the critical angle-of-attack. A stall buffet can range from hardly noticeable, to significant shaking. See *Stall*.

Stall Recovery Maneuver a pilot can fly to lower the angle-of-attack below the critical angle-of-attack, and resume good airflow over the wings (i.e., to resume flying).

Stall Speed The speed at which a plane is flying when the angle-of-attack becomes too great and the stall occurs. For a given aircraft, stalls always occur at the same angle-of-attack, but stall speed can vary with a plane's weight and location of its center of gravity, among other things.

Steam Gauges Analogue gauges, as opposed to a glass cockpit. See *Glass Cockpit* and *Six-pack*

Sterile Cockpit Describes a cockpit adhering to the FAA Sterile Cockpit Rule, that when flying below 10,000 feet and during a critical phase of flight (such as takeoff and landing) pilots may only speak about matters pertaining directly to the safety of their flight.

Stick See *Joystick* and *Sidestick*.

Stick-and-Rudder Skills A reference to a pilot's ability to hand-fly a plane well. For instance, one might be described as a natural stick-and-rudder pilot.

Stick Shaker A warning device in some aircraft to alert the pilot a stall is imminent. When triggered it causes the control column and yoke to vibrate and produces a loud rattling noise.

Threshold The start of a runway.

TOGA TakeOff/Go Around. Full-forward throttle setting for maximum engine thrust.

Trailing Edge The rear edge of a wing.

Trimotor A three-engine propeller plane. A popular design in the 1920s and 1930s.

Trim Wheel A wheel in the cockpit, mounted perpendicular to the instrument panel by the pilot's thigh, connected to the horizontal stabilizer. Moving the wheel moves the leading edge of the horizontal stabilizer up and down, to raise and lower the nose in flight. The horizontal stabilizer will remain in whatever position the trim wheel moves it.

Tunneling A conscious state where one is intensely and narrowly focused on a single task, visual stimulus, or data source (such as a cockpit instrument), to the near-total or even complete exclusion of everything else.

Type Rating An FAA-issued rating signifying legal permission to fly a certain specific type of aircraft, such as a Gulfstream private jet, or a Boeing 737.

VASI Visual Approach Slope Indicator lights. Two sets of lights, one behind the other, on the left side of a runway, visible to a pilot on approach, that change color based on whether a plane is above, on, or below the glide slope. It usually contains two sets of three lights each, although it can be a set of one or two lights each as well. While flying the approach, the lights appear to the pilot to be one above the other, and are red and white. The light sequence is: both light sets white = too high; top set red, bottom set white = on the glide slope; both sets red = too low. See *PAPI.*

Vertical stabilizer The tall vertical structure in a plane's tail that helps give it latitudinal (left-and-right) stability. The rudder is attached to the rear of the horizontal stabilizer.

Visual Approach An approach conducted under Instrument Flight Rules but where the runway is visible from a few miles away and a few thousand feet altitude. The pilot guides the airplane to the runway using visual cues and not instruments (although instruments may be used).

Visual Flight Rules (VFR) The conditions for flying where a pilot uses visual references outside the cockpit to control the airplane. When flying under VFR, weather conditions must be good enough so that a pilot can navigate and land without resorting to using instruments. Certain specific weather conditions do not allow for VFR flight, including cloud bases below a minimum height above the ground, and visibility below a minimum distance.

VFR Sectional Map A map of a segment of the country, containing everything a pilot would need to know when planning and flying a trip under Visual Flight Rules. Sectionals contain airport and navigation frequencies, obstacles such as mountains, buildings and antenna towers, runways, restricted-flight areas (for instance, areas that are part of major airport operations, and military practice and en-route areas), landmarks such as highways, train tracks, lakes and rivers, and other important information.

Waypoint A geographic point on a map used for navigation purposes. It could be an intersection of radio beams, the source of a radio navigation beacon, or a point identified only by latitude and longitude on GPS. Names are often gibberish, occasionally meaningful to the area, and sometimes funny. For example, Lebron James, the basketball player, has one near Cleveland named for him: LEBRN, pronounced like his name.

Yaw A sideways motion – left or right – of a plane's nose.

Yoke A half-wheel used by a pilot to control a plane. Mounted either directly into the instrument panel, or on a column in front of the pilot. To bank the plane left or right, the yoke is turned in the desired direction. Pulling the yoke makes a plane climb. Pushing it makes it descend.

Sources

Many of the details provided in this book are common knowledge, easily verifiable and widely known, and therefore not cited. However, where I feel citing is required, my sources are below. Generally, all information regarding an aircraft accident, including cockpit voice recordings, was taken from its official report produced by a national aviation authority, and cited as such at the beginning of the book's review and discussion of that accident.

Chapter 1. February 12, 2009

Details of Colgan flight 3407 from: National Transportation Safety Board report, *Loss of Control on Approach Colgan Air, Inc. Operating as Continental Connection Flight 3407 Bombardier DHC-8-400 N200WQ Clarence Center, New York February 2, 2009.* AAR-10/01 PB2010-910401. Washington, DC, February 2, 2010.

Chapter 2. Born to Fly

It took another ten years: 'A Brief History of Aerobatics.' *Airtug.com*, December 14, 2016.

A small minority of commercial airline pilots: Of all the pilots I've known, only a handful didn't want to fly their entire lives. Also Bean, Larry, 'Robb Report Pilots Survey: Everything You Wanted to Know About the People Flying Your Plane.' *Robb Report.* November 23, 2017; robbreport. com.

Chapter 3. Human Factors

Multiple studies confirm it: One study pertaining to aviation is found in: Fanjoy, Richard, Julius Keller, Julius. 'Flight Skill Proficiency Issues in Instrument Approach Accidents.' *Journal of Aviation Technology and Engineering* 3:1 17–23 (2013). This documents and researches pilot proficiency in hand-flying instrument approaches, which are almost always flown on autopilot these days. When pilots do not have the benefit of autopilot, their ability to fly the approach is often severely lacking.

Chapter 5. Elmer Sperry

Much of Elmer Sperry's biographical history, including background on Richard Sperry, from: Hunsaker, J.C. *Elmer Ambrose Sperry 1860-1930 A Biographical Memoir*. Washington, DC, National Academy of Sciences. 1954. Also 'Elmer Sperry Dies; Famous Inventor.' New York, *New York Times*, June 17, 1930. Also 'Elmer Ambrose Sperry Sr.' *Wikipedia*.

Details of the gyrocompass lawsuit between Elmer Sperry and Hermann Anschütz-Kaempfe from: Bray, Hiawatha. *You are Here: From the Compass to GPS, the History and Future of How We Find Ourselves*. Philadelphia, PA Basic Books, 2014. Also Galison, Peter, *How Experiments End*, Chicago, IL, University of Chicago Press. 1987.

Chapter 6. Lawrence Sperry

Details of Lawrence Sperry's life from: Scheck, William, 'Lawrence Sperry: Genius on Autopilot.' *Aviation History Magazine*, November, 2004 (from *historynet.com*). Also Davenport, William W., *Gyro! The Life and Times of Lawrence Sperry*. New York, NY, Charles Scribner's Sons, 1978.

Additional details of Lawrence Sperry's and Dorothy Peirce's crash in Great South Bay off Long Island from: 'From Her Sick Bed Plans New Flights.' *New York Times*, November 28, 1916. The article misspells Ms. Peirce's last name. Also 'Someone Had to be First...' *Check-Six.com*.

Details of Lawrence Sperry's demonstration flights from: previously cited article by William Scheck and book by William Davenport, and from: Ide, John Jay. 'The Sperry Gyroscopic Stabilizer: How it is Constructed, How it Operates, and How it Demonstrated its Capabilities During an Interesting Test in France.' *Scientific American*, August 8, 1914, page 96. Also from Hoeber, Harold, 'Long-Sought Aeroplane Stabilizer Invented.' *New York Times*, July 19, 1914 (*aero*plane in the headline is not a typo).

Chapter 7. Lost Decades

Details of Lawrence Sperry's death from: Davenport, *Gyro!*

Details of Wily Post's flight from: 'Post Arrives Safely in New York, Circling the World in 7 Days, 19 Hours; Mollisons are Flying the Atlantic.' *New York Times,* July 23, 1933.

SOURCES

Chapter 11. Autos
Around half of all fatal commercial plane crashes: 'Statistical Summary of Commercial Jet Airplane Accidents 1959-2017,' published by *Boeing Commercial Airplanes*, October 2018.

Chapter 13. Sniffles
Details of Captain Marvin Renslow from: Altman, Howard, 'In New York, a Plane Crash; at Home, Grief.' *Tampa Tribune*, February 15, 2009. Also from *findagrave.com/memorial/marvin-renslow*. Also from National Transportation Safety Board report on the crash of Colgan 3047, *Loss of Control on Approach*.

Details of First Officer Rebecca Shaw from Miletich, Steve and Broom, Jack. 'Friends, Family Mourn Loss of Copilot from Washington.' *Seattle Times*, February 14, 2009.

Chapter 17. The Situation
Conjecture that Oswald Boelke first introduced the concept of situational awareness from: Stanton, N.A., Chambers, P.R.G., Piggott, J., 'Situational Awareness and Safety.' *Safety Science 39*, pp. 189-204 (2001).

Raja Parasuraman's biographical information from: Brookhuis, Karel, 'In Memoriam Raja Parasuraman.' *Int. J. Human Factors and Ergonomics, Vol. 3, Nos. 3/4, 2015* 227, found on *inderscience.com*, which was reprinted with permission from a newsletter dated 1/2015 of the Europe Chapter of the Human Factors and Ergonomics Society. Also from *dignitymemorial.com/obituaries/falls-church-va*.

The definition of automation from: Parasuraman, Raja, Sheridan, Thomas, and Wickens, Christopher, 'A Model for Types and Levels of Human Interaction with Automation.' *IEEE Transactions on Systems, Man, and Cybernetics – Part A: Systems and Humans,* vol. 30, no. 3 (May 2000). That piece cites another article by Parasuraman, with Riley, V.A., entitled 'Humans and Automation: Use, Misuse, Disuse, Abuse.' *Human Factors.* vol. 39, pp. 230-53. Also from, Sheridan, Thomas, Parasuraman, Raja. '*Human Automation Interaction.*' June 1, 2005.

Merriam-Webster.com contains this definition: Self-satisfaction especially when accompanied by unawareness of actual dangers or deficiencies.

Chapter 18. Hazardous States
'I discovered,' he said: in a letter from Alan Pope to the author, January 20, 2020.

His research culminated in a report: Pope, Alan T. and Bogart, Edward H. 'Identification of Hazardous Awareness States in Monitoring Environments,' SAE Transaction Vol 101, Section 1, *Journal of Aerospace* (1992).

Prinzel's constructs from: National Aeronautics and Space Administration. *Research on Hazardous States of Awareness and Psychological Factors in Aerospace Operations*. Lawrence J. Prinzel III, NASA/TM-2002-211444. NASA Langley Research Center, Hampton, VA, March 2002.

Chapter 19. Engine Gauges

Raja Parasuraman's 1993 research from: Parasuraman, Raja, Molloy, Robert, Singh, Indramani, 'Performance Consequences of Automation Induced Complacency.' *International Journal of Aviation Psychology*, (February 1993).

Ray Comstock's MATB from: National Aeronautics and Space Administration. *Multi-Attribute Task Battery for Human Operator Workload and Strategic Behavior Research*. J.R. Comstock and R.J. Arnegard. Tech Memorandum No. 104174, NASA Langley Research Center, Hampton, VA 1992. Abbreviation and pronunciation 'MATB' and 'mat-bee' from: letter from Comstock to Alan Pope, April 10, 2020.

Chapter 20. San Francisco

Details of Asiana 214, as well as the cockpit CVR, from: National Transportation Safety Board, *Descent Below Visual Glidepath and Impact with Seawall Asiana Airlines Flight 214 Boeing 777-200ER HL-7742 San Francisco California July 6, 20013*, NTSB/AAR 14/01 PB214-105984, Washington, DC, June 24, 2014.

Chapter 24. Bias and Surprise

The computerized flight planning study from: Layton, C., Smith, P.J., & McCoy, C.E. 'Design of Cooperative Problem-solving System for En-Route Flight Planning: an Empirical Evaluation.' *Human Factors*, 36, (1994). A summary of this study is provided in Parasuraman, Raja and Manzey, Dietrich, 'Complacency and Bias in Human Use of Automation: An Attentional Integration.' *The Journal of Human Factors and Ergonomics Society*, June 2010.

Chapter 25. LIFUS

Details of Turkish Airlines flight 1951 from: The Dutch Safety Board. *Crashed During Approach, Boeing 737-800, Near Amsterdam Schiphol Airport, 25 February 2009*. Project number M2009LV0225_01 The Hague, May 2010.

SOURCES

Chapter 27. Perfect Storm
A textbook definition: Robert Mauro, 'Detecting and Mitigating Automation Surprise.' *Decisionresearch.org* a website of Decision Science Research Institute, Inc.

Chapter 28. Airliner
Flying commercially is statistically: Bureau of Transportation Statistics, US Department of Transportation, *bts.gov/archive/publications/by_the_numbers/transportation_safety.*

Igor Sikorsky's background and aircraft details from: Richard Hallion, 'Airplanes that Transformed Aviation.' *Airspacemag.com.* Also Sikorskyarchives.com/history. Also *wikipedia.com.*

Details of Knute Rockne's death from: 'Farmer Tells Story.' *Spokane Daily Chronicle*, March 31, 1931.

Chapter 29. Boeing
Details of the Boeing 247 from: 'Boeing Model 247.' *The Aviation History Museum Online.* Aviation-history.com. Also from: 'Boeing's new Model 247 Transport.' *Aviation Week & Space Technology*, April 1933.

Details of the DC-3 from: 'Douglas DC3.' *The Aviation History Museum Online.* Aviation-history.com.

Chapter 31. Airbus
The memorandum quote, market share data, as well as other details of Airbus's history was found on the Airbus corporate website, *airbus.com (http://www.airbus.com/company/history/the-narrative/).*

As far as I understand it: 'Eastern Leases Four A300Bs.' *Aviation Week & Space Technology*, May 9, 1977, pp. 22-23.

Chapter 32. Fly-by-Wire
In 1981 Airbus finalized: from Airbus corporate website, *airbus.com.*

Ziegler and Airbus made: from Airbus corporate website. Also from 'Farnborough: Airbus's fly-by-wire pioneered Bernard Ziegler wins Flightglobal Lifetime Achievement Award,' *flightglobal.com*, July 11, 2012.

The customer acceptance: 'Airbus vs. Boeing, the Challenger Gains Ground.' *Aviation Week and Space Technology*, January 10, 2000, p. 45.

Chapter 33. Guardrails
The biggest improvement: Untitled, *Aviation Week and Space Technology*, September 8, 1986, p. 42.

The guardrail is not there: 'Airbus May Add to A320 Safeguards, Act to Counter Crew "Overconfidence,"' *Aviation Week and Space Technology*, April 30, 1990, p. 59.

The 1999 Airline Pilots Association test is documented in: 'Flight Test Results of the Controlled Flight Into Terrain Avoidance Maneuver in Fly-By- Wire Transports,' 1999, by the Airline Pilots Association, Airworthiness Performance Evaluation and Certification Committee, Captain Ron Rogers, Chairman.

He is famously quoted as saying: Langewiesche, William. 'The Human Factor.' *Vanity Fair*, September 17, 2014 (October issue). Also Langewiesche, 'What Really Brought Down the Boeing 737 Max,' *New York Times*, September 18, 2019.

Chapter 34. neo

Cost more than $32 billion: Gates, Dominic, 'Boeing Celebrates 787 Delivery as Program's Costs Top $32bb.' *Seattle Times*. September 24, 2011.

Chapter 35. Runaway Trim

Boeing engineers first tried: '737 Max – MCAS' *The 737 Information Site: b737.org.uk*

With the autopilot engaged: Peter A. Bedell. 'Opinion: Lessons from the 737 Max Debacle; Know How to Disengage Any System that can Grab Your Controls.' *AOPA Website aopa.org*. August 1, 2019.

Boeing supposedly agreed to rebate: Paszor, Andy and Tangel, Andrew. 'Ex-Boeing Pilot Complained of Management Pressure on MAX, Former Colleagues Say.' *Wall St. Journal*, October 23, 2019.

Boeing's expectation that pilots would interpret MCAS problems as runaway trim was reported by Boeing and is common knowledge, though that, as well as the American Airlines statistic, are both from: McCartney, Scott. 'Inside the Effort to Fix the Troubled Boeing 737 Max.' *Wall Street Journal*, June 5, 2019.

Chapter 37. MCAS

Details of Lion 610 from the accident report by Indonesian transportation authorities: Komite Nasional Keselamatan Transportasi (KNKT). *Aircraft Accident Investigation Report PT. Lion Mentari Airlines Boeing 737-8 (MAX); PK-LQP Tanjung Karawang, West Java Republic of Indonesia*. Final KNKT.18.10.35.04 Jakarta, Indonesia. 29 October 2018. The preliminary report dated November 2018 was also used. The reports do not contain a CVR transcript, though it does include occasional words in quotes.

Chapter 38. Ethiopian Air

Details of Ethiopian 302 from: Federal Democratic Republic of Ethiopia Ministry of Transport Aircraft Accident Investigation Bureau. *Aircraft Accident Investigation Preliminary Report, Ethiopian Airlines Group B737-8 (MAX) Registered ET-AVJ* AI-01/19 March 10, 2019. Also from: Federal Democratic Republic of Ethiopia Ministry of Transport Aircraft Accident Investigation Bureau. *Interim Investigation Report on Accident to the B737-8 (MAX) Registered ET-AVJ* operated by Ethiopian Airlines on 10 March 2019 Report No. AI-01/19, March 9, 2020.

Getachew's and Mohammed's backgrounds: 'Youngest Captain, Loving Son: Ethiopian Pilots Honored in Death.' *Reuters*, March 20, 2019.

Chapter 39. ADIRU

Three was a 'nonstarter': Nicas, Jack and Creswell, Julie, 'Boeing's 737 Max: 1960s Design, 1990s Computer Paper and Paper Manuals. *New York Times* April 8, 2019.

Information on Lion Air flight 1043 is from the KNKT Final Report on the crash of Lion Air Flight 610, October 2018. The report does not say how the captain of 1043 knew his primary flight display was the incorrect one, only suggesting his actions 'may have indicated that the Captain generally was aware of the repetitive previous problem' on the previous flights.

Chapter 40. Captain's Choice

The description of pilots from Licht, Bill, 'Discussing the Pilot Personality,' *Air Line Pilot*, June-July 2016. Also in Weiss, Vern, 'Is There Such a Thing as a Pilot Personality.' *DisciplesofFlight.com*, March 26, 2016.

News reports said: There were a number of reports, but one is: 'I'm Coming Home Pilot Told Mother Before Ethiopian Airline Crash.' Allafrica.com. March 13, 2019.

Chapter 41. Airmanship

A Boeing employee confirmed it: Gelles, David, 'Crisis Reveals "Sick" Culture at Boeing.' *New York Times*. January 11, 2020.

Since we are on the topic of statistics: The National Safety Council. *Nsc. org*. Also Mazzei, Patricia, 'Opioids, Car Crashes and Falling: The Odds of Dying in the U.S.' *New York Times*, January 14, 2019.

There are 290,000: 'Airline Pilot Demand Outlook 10 Year View.' Published by CAE, Inc, June 2017.

Chapter 42. South Atlantic

Details of Air France flight 447 from: the English language version of: Bureau of Enquiry and Analysis for Civil Aviation Safety (Bureau d'Enquêtes et d'Analyses pour la sécurité de l'aviation civile). *Final Report on the Accident on 1st June 2009 to the Airbus A330-203 registered F-GZCP operated by Air France flight AF 447 Rio de Janeiro – Paris*, Le Bourget Cedex, France, July 2012. The English version of Air France 447's cockpit conversation is from Appendix 1 of this report. Also from Palmer, Bill, Understanding Air France 447, self-published 2013.

Chapter 43. Airspeed

Bonin pulled his sidestick back: Control input details from: Bureau d'Enquêtes et d'Analyses pour la sécurité de l'aviation civile. *Animation. Accident on June 1st 2009 to the Airbus A330-203 registered F-GZCP operated by Air France flight AF 447 Rio de Janeiro–Paris*. Video produced by BEA and posted December 12, 2014. The video shows control inputs, the captain's primary flight display, throttle settings, and the plane's track for the flight during the period from just before the pitot tubes failed, through the crash.

Chapter 45. Hippocratic Oath

They are not the first crew: Fiorino, Frances, 'Hidden Dangers in the Cockpit.' *Aviation Week & Space Technology*. July 6, 2009.

Chapter 46. ECAM

Details of Qantas 32 from: Australian Transport Safety Bureau. *In-flight Uncontained Engine Failure Airbus A380-842 VH-OQA Overhead Batam Island, Indonesia, 4 November, 2010*. AO-2010-089 Final, Canberra, Australia, 2013. Also from Crespigny, Richard de, *QF32 The Captain's Extraordinary Account of How One of the World's Worst Air Disasters Was Averted*, Sydney, Australia, Pan Macmillan Australia Pty Ltd. 2012.

Chapter 47. Seatbelts

Details of Qantas 72 from: Australian Transport Safety Bureau, *In-flight Upset 154 km West of Learmonth, WA, 7 October 2008 VH-QPA Airbus A330-303*, A02008-070 Final, Canberra, Australia, 2011.

Chapter 48. Pan-Pan

Broken his nose, and *lacerations and broken bones*: O'Sullivan, Matt, 'The Untold Story of QF72: What Happens When 'Psycho' Automation Leaves Pilots Powerless?' *Sydney Morning Herald*, May 12, 2017.

SOURCES

Chapter 50. Distraction
Details of the Miami-bound L1011 from: National Transportation Safety Board. *Aircraft Accident Report, Eastern Airlines Inc., L1011, N310EA Miami, FL.* Washington, DC, June 14, 1973.

A study by human factors experts: Dismukes, Key, Young, Grant, and Sumwalt, Robert, 'Cockpit Interruptions and Distractions: Effective Management Requires a Careful Balancing Act.' November 1998.

True multitasking is a myth: A video monologue by Captain Sullenberger. *Captain Sully's Minute-by-Minute Description of the Miracle on the Hudson*, produced by *inc.com*, posted on YouTube March 6, 2019.

Chapter 51. Canada Goose
Details of US Airways 1549 from: National Transportation Safety Board, *Aircraft Accident Report Loss of Thrust in Both Engines After Encountering a Flock of Birds and Subsequent Ditching on the Hudson River US Airways Flight 1549 Airbus A320-214, N106US Weehawken, New Jersey January 15, 2009.* NTSB/AAR-10/03 PB2010-910403, Washington, DC. Also from a video animation, including the voice recordings from the CVR, produced by the NTSB. Also from *Minute-by-minute...* video monologue by Captain Sullenberger.

Aimed straight north: The northerly heading was given by air traffic controller Patrick Harten in prepared testimony before the US Congress Transportation and Infrastructure Subcommittee on Aviation, February 24, 2009. NATCA.org (the National Air Traffic Controllers Association website).

BRACE BRACE BRACE: From *Minute-by-minute...* video monologue by Captain Sullenberger.

Wrap my mind around: Patrick Harten, before the US Congress, February 24, 2009.

Well, that wasn't as bad: From *Minute-by-minute...* video monologue by Captain Sullenberger.

Chapter 53. And Furthermore
Humans have the capability: Dubois, Thierry, 'Research on Brain Activity to Help Cockpit Design.' *Aviation Week & Space Technology*, January 27, 2020.

Costing Boeing more than: Cameron, Doug and Tangel, Andrew. 'Boeing Posts Full-Year Loss Amid 737 Max Setbacks.' *Wall Street Journal*, January 29, 2020.

We have always favored: Norris, Guy, 'Clean Sheet.' *Aviation Week & Space Technology*, February 10-23 2020, pp. 26-28.

Additional Sources

Bhana, Hemant, 'Trust but Verify.' *Aerosafety World*, June 2010.

Federal Aviation Administration Audit Report, Enhanced FAA Oversight Could Reduce Hazards Associated with Increased Use of Flight Deck Automation, Matthew Hampton, AV-2016-013Washington, DC, 2016.

Gawron, Valerie. Nothing Can Go Wrong – A review of Automation-Induced Complacency Research, MITRE Corp, McLean, Va, 2019.

Hiraki J., Warnink M. 'Cockpit Automation Fact Sheet.' *Aviationfacts.eu.* 2016.

Koeppen, Nicholas A. *The Influence of Automation on Aviation Accident Fatality Rates: 2000-2010*, Embry-Riddle Aeronautical University, 2012.

National Aeronautics and Space Administration, *Examination of Automation-Induced Complacency and Individual Difference Variates*, Lawrence J. Prinzel III, NASA Langley Research Center, Hampton, VA, December 2001.

Newhouse, John, *Boeing versus Airbus*, New York, New York, Alfred A. Knopf Inc, 2007.

Newhouse, John, *The Sporty Game*, Knopf, 1982.

Pope, Alan T., Stephens, Chad L., Scerbo, Mark W. *Adaptive Automation for Mitigation of Hazardous States of Awareness*, 2011.

Index